Complexity and the Human Experience

Complexity and the Human Experience

Modeling Complexity in the Humanities and Social Sciences

edited by
Paul A. Youngman | Mirsad Hadzikadic

PAN STANFORD PUBLISHING

Published by

Pan Stanford Publishing Pte. Ltd.
Penthouse Level, Suntec Tower 3
8 Temasek Boulevard
Singapore 038988

Email: editorial@panstanford.com
Web: www.panstanford.com

British Library Cataloguing-in-Publication Data
A catalogue record for this book is available from the British Library.

Complexity and the Human Experience: Modeling Complexity in the Humanities and Social Sciences

ISBN 978-981-4463-26-3 (Hardcover)
ISBN 978-981-4463-27-0 (eBook)

Printed in the USA

Contents

Preface

In an attempt to capture the back and forth between chaos and order, Mitchell Waldrop, author of *Complexity* (Simon & Schuster, 1992), cites the computer scientist Christopher G. Langton's coining of the phrase "edge of chaos." Langton uses this phrase to denote that mysterious "something" that makes life and mind possible. He posits that the edge of chaos exists in that interplay between order and disorder that is characteristic of all complex systems, be they economies, minds, or bodies. In scientific terms for example, in the molecule composed of two parts hydrogen and one part oxygen, the "edge of chaos" is that state between ice where atoms are locked into place and water where they randomly roil.

This volume represents an effort to expand the edge, so to speak. We seek to find the edge of chaos in fields not generally associated with complex adaptive systems. Is there, for example, an interaction between order and disorder in a work of literary fiction? And, more important, is it identifiable in quantitative terms? In other words, this volume attempts to apply the principles of complexity science to new fields. We therefore offer an array of essays from the social sciences and the humanities that either present or analyze a computer model to approach traditional questions their respective fields in an exciting, new way. In this respect, our volume simultaneously breaks new ground and creates new ground.

As the editors of this collection, we are grateful to our authors for their willingness to risk subjecting their traditional research questions in the social sciences and the humanities to this novel approach. We are also grateful to the Office of Digital Humanities at the National Endowment for the Humanities for sponsoring the "Institute for Advanced Topics in the Humanities: Computer Simulations in the Humanities" at the University of North Carolina Charlotte in 2011. Many of our authors were attendees. It is therefore our great hope that this volume represents a continuation of that very stimulating first conversation.

Paul A. Youngman
Mirsad Hadzikadic
Lexington, VA, and Charlotte, NC
2014

Introduction: Computer Modeling in Humanities and Social Sciences

Paul A. Youngman[a] and Mirsad Hadzikadic[b]

[a]*Department of German and Russian, Washington and Lee University, 204 W. Washington St., Lexington VA, 24450, USA*
[b]*University of North Carolina Charlotte, 9201 University City Blvd., Charlotte, NC 28223, USA*

youngmanp@wlu.edu, mirsad@uncc.edu

Complexity science is the study of how large numbers of relatively simple entities organize themselves into a collective whole that creates patterns, uses information, and, in some cases, evolves and learns. Those collective wholes that do not evolve and learn are complex systems; those that do are complex adaptive systems (CAS). Complexity and its various systems have been a topic of study in the natural sciences for decades already (Mitchell, 2011). Physics, chemistry, biology, mathematics, meteorology, and engineering practitioners have used the concept of complex systems to explain phenomena as diverse as phase transitions in physical matter, immune system functions, and weather patterns (Braha et al., 2006; Callebaut and Rasskin-Gutman, 2005; Flake, 1998; Gell-Mann, 1994; Gleick, 1988; Nicolis and Prigogine, 1989; von Bertalanffy, 1969). The CAS these scientists have modeled use, for the most part, the concept of dynamic systems and nonlinear equations. In short, CAS is a nice fit for the natural sciences in many ways. The goal of this volume, however, is to show where and how the idea of complexity has spread beyond the natural sciences. Our authors show how complexity ontology with its corresponding emphasis on modeling has already effectively spread to the social sciences and is at the very threshold of making a significant impact on the humanities.

Part 1 of our volume is, therefore, devoted to the humanistic threshold. In this research realm, the effective use of CAS is still in the early stages. There are many obstacles to overcome in the humanities before such an approach becomes more widespread. As one conference attendee quipped, "If you think that agents in the social sciences have 'will,' try the humanities." Due to its computational nature, CAS requires one to think of the object of one's study as data and thereby as quantifiable. The humanities, on the other hand, involve the study of the human in all of its imprecision, its subjectivity, and its non-quantifiability, using methods that are critical, interpretive, and even speculative. CAS smacks of precision, objectivity, and quantifiability, but like all sciences, it, too, involves a great deal of judgment, informed guesswork, and approximations. Moreover, CAS most certainly understands the imprecision of the human, especially when more than one interact together. Given the difficulty of the human, CAS understands above all that its models are statistical approximations of what, in fact, may be. Just as a novel or a painting is an aesthetic approximation of what may be.

While many humanities scholars would not be comfortable with CAS for the myriad reasons listed above, this volume is a testament to the growing importance of such techniques in humanities research. The diversity of scholars also demonstrates broad suitability in the humanities and not just in isolated fields. The first chapter of Part 1 of this volume is from the philosopher Patrick Grim and a team of co-authors. "Philosophical Analysis in Modeling Polarization: Notes from a Work in Progress" provides an analysis of belief polarization in U.S. society using agent-based modeling. Among other things, Grim and his co-authors second the editors' contention that the modeling aspect of CAS helps in conceptual understanding because it forces the researcher to clarify core concepts. Chapter 2 is a contribution from Markus Christen titled "Overcoming Moral Hypocrisy in a Virtual Society." This work uses agent-based modeling to understand the effect of various social strategies on the prevalence of moral hypocrisy in a virtual society. Chapter 3 by Rory Smead is titled "Evolving Games and the Social Contract." Smead deals with modeling evolving games using existing tools in evolutionary game theory and then modeling simple case studies relevant to the formation or dissolution of social contracts. Graham Sack's "Character

Networks for Narrative Generation" serves as Chapter 4 to this volume. Sack's contribution applies social network analysis to narrative generation with an eye toward supplementing existing methods used by artificial intelligence researchers. Chapter 5 is a contribution from Jeremy Throne. "Modeling the Communications Circuit: An Agent-Based Approach to Reading in 'N-Dimensions'" uses complexity theory and agent-based modeling to show that literary circulation is not an orchestrated dynamic between the agents known as author, editor, social structure and economic system, but rather emergent behavior due to the uncontrolled interaction between those agents. The historian Marten Düring contributes the final chapter of Part 1. "The Potential of Agent-Based Modeling for Historical Research" is an overview of the successful application of agent-based modeling by historians with an emphasis on counterfactuals in history.

In the past 10 to 15 years, social scientists have begun experimenting with the complex systems tools developed in physics, mathematics, and engineering in exciting ways in order to better understand the nature of key issues facing society today. One important difference between social systems and the natural systems that were the focus of CAS in the early years of its development is that social systems involve the active participation of system elements. In a sense, the agents in the systems that are the focus of social scientists possess "will," thereby making the CAS challenge to the social sciences that much greater, especially when it comes to modeling. Despite this, social scientists have made tremendous progress in applying the methodology of CAS to a diverse number of fields ranging from economics, sociology, transportation studies, warfare studies, decision making, and many other disciplines. Already in the 1990s, M. Resnick was exploring the parallel worlds of turtles, termites, and traffic jams using the CAS paradigm (Resnick, 1994). Some of the most notable projects in the twenty-first century include G. G. Marten's use of self-organization and emergent properties in CAS to understand and explain human–ecosystem interaction (Marten, 2001); R. K. Sawyer's simulation of social emergence with artificial, agent-based societies, which led him to define societies themselves as complex systems (Sawyer, 2005); Pascale, Milleman, and Gioja's work that defines CAS as the very law of nature and suggests that CAS should therefore become the new law of business (Pascale et al.,

2000); M. Buchanan's use of CAS principles to explain why large-scale catastrophes happen (Buchanan, 2002); M. Farnsworth's use of CAS to better explain the water-cycle in the systems of Roman culture (Farnsworth, 2001); finally, N. E. Harrison's application of CAS to provide a unique understanding of complexity in comparative world politics (Harrison, 2006).

Part 2 of our volume builds on this foundational social science research in CAS. It highlights the work of a diverse group of social scientists who deploy CAS to a variety of ends. In Chapter 7, Martin Zwick leads off the social science section with a contribution titled "Complexity Theory and Political Change: Talcott Parsons Occupies Wall Street." His contribution combines polity, economy, community, and culture and applies complexity theory to Talcott Parsons' model of societal structure to analyze the recent Occupy Wall Street movement. In Chapter 8, "China's Complex Policy Network," Jerome Sibayan maps the policy network that is the Central Committee of the Communist Party in China. In doing so, he shows how the network exhibits the complex characteristics of small worlds, emergent order, emergent functionality, and resilience. Chapter 9, Dana Downey's "Convergence versus Emergence of Youth Extremism: An Agent-Based Model of the Arab Spring," is an analysis of patterns of convergence and emergence and their role in the radicalization of an epistemic network of the Arab Spring. Russell S. Gonnering and his team of co-authors contribute the tenth chapter titled "Complexity of Structure in Electronic Medical Record Systems"—an investigation into the complexity of structure in three Electronic Medical Records (EMR) in order to contrast their findings with those in transcribed paper records, dictated reports, and articles in biomedical journals. In Chapter 11, "Adaptation of Foodways and Indicators of Malnutrition: Philippines as a Case Study," Amanda Andrei employs power law analysis to provide insights into managing food security and stability. Chapter 12 includes an archaeological contribution from Thaer W. Jayyousi and Robert G. Reynolds. "Exploiting the Synergy between Micro, Meso, and Macro Levels in a Complex System: Bringing to Life an Ancient Urban Center" constructs agent-based models using computational intelligence techniques, complex systems, data mining, as well as existing archaeological data from the prehistoric urban center, Monte Albán, in order to reconcile the temporal and spatial expanse of the site. The thirteenth and final chapter is a contribution by Michael Givel and Liz Johnson. "Scientific Paradigms in U.S. Policy: Is It Time for Complexity Science?" is a discussion of the shortcomings of U.S. public policy development and an analysis of complexity science as an answer to some of those shortcomings.

This volume is the outcome of two events. The first was a 2011 Summer Institute for Advanced Topics in the Digital Humanities sponsored by the Office of Digital Humanities at the National Endowment for the Humanities. The Institute, Computer Simulations in the Humanities, was held at the University of North Carolina Charlotte and was host to an eclectic group of scholars from various social science and humanities fields as well as from many countries around the world. Building on the momentum of the Summer Institute, the North Carolina Complex Systems Institute, the Center for Advanced Research in the Humanities, the College of Computing and Informatics, and the College of Liberal Arts and Sciences all at UNC Charlotte hosted the 1st Annual Conference on Complexity and Human Experience in May 2012—an energetic gathering that may have surpassed the Summer Institute in diversity of fields. The vast majority of the chapters in this volume were first developed at the Summer Institute and presented at the Conference. We are especially grateful to Ted Carmichael for his help in organizing the Conference and to Marvin Croy, Tony Beavers, and Patrick Grim, who organized the Summer Institute.

Bibliography

Axelrod, R. M., and Cohen, M. D. 1999, *Harnessing Complexity: Organizational Implications of a Scientific Frontier*, Free Press, New York.

Bonabeau, E., Dorigo, M., and Theraulaz, G. 1999, *Swarm Intelligence: From Natural to Artificial Systems*, Oxford University Press, New York.

Braha, D., Minai, A. A., and Bar-Yam, Y. 2006, *Complex Engineered Systems: Science Meets Technology*, Springer, Berlin; New York.

Buchanan, M. 2002, *Ubiquity: Why Catastrophes Happen*, Three Rivers Press.

Callebaut, W., and Rasskin-Gutman, D. 2005, *Modularity: Understanding the Development and Evolution of Natural Complex Systems*, MIT Press, Cambridge, Mass.

Capra, F. 1982, *The Turning Point: Science, Society, and the Rising Culture*, Simon and Schuster, New York.

Capra, F. 2002, *The Hidden Connections: Integrating the Biological, Cognitive, and Social Dimensions of Life into a Science of Sustainability*, 1st ed., Doubleday, New York.

Dooley, K. J. 1997, A complex adaptive systems model of organization change, *Nonlinear Dynamics, Psychology, and Life Sciences*, **1**(1), pp. 69–97.

Durlauf, S. N., and Young, H. P. 2001, *Social Dynamics*, Brookings Institution Press; MIT Press, Washington, D. C.; Cambridge, Mass.

Epstein, J. M. 2006, *Generative Social Science: Studies in Agent-Based Computational Modeling*, Princeton University Press, Princeton.

Farnsworth, R. 2001, *Mediating Order and Chaos: The Water-Cycle in the Complex Adaptive Systems of Romantic Culture*, Rodopi, Amsterdam; New York.

Flake, G. W. 1998, *The Computational Beauty of Nature: Computer Explorations of Fractals, Chaos, Complex Systems, and Adaptation*, Cambridge, Mass: MIT Press, c1998.

Gell-Mann, M. 1994, *The Quark and the Jaguar: Adventures in the Simple and the Complex*, W. H. Freeman and Co., New York.

Gell-Mann, M. 1995, Complex adaptive systems, *The Mind, The Brain, and Complex Adaptive Systems*, pp. 11–24.

Gleick, J. 1988, *Chaos: Making a New Science*, Penguin, New York, N.Y., U.S.A.

Gribbin, J. R. 2004, *Deep Simplicity: Bringing Order to Chaos and Complexity*, 1st U. S. ed., Random House, New York.

Harrison, N. E. 2006, *Complexity in World Politics: Concepts and Methods of a New Paradigm*, State University of New York Press, Albany.

Hazy, J. K., Goldstein, J. A., and Lichtenstein, B. B. (eds.) 2007, *Complex Systems Leadership Theory: New Perspectives from Complexity Science on Social and Organizational Effectiveness (Exploring Organizational Complexity)*, ISCE Publishing.

Holland, J. H. 1992, *Adaptation in Natural and Artificial Systems: An Introductory Analysis with Applications to Biology, Control, and Artificial Intelligence*, 1st MIT Press ed., MIT Press, Cambridge, Mass.

Holland, J. H. 1995, *Hidden Order: How Adaptation Builds Complexity*, Addison-Wesley, Reading, Mass.

Holland, J. H. 1998, *Emergence: From Chaos to Order*, Addison-Wesley, Reading, Mass.

Innes, J. E., and Booher, D. E. 1999, Consensus building and complex adaptive systems, *Journal of the American Planning Association*, **65**(4), pp. 412–423.

Johnson, N. F. 2007, *Two's Company, Three Is complexity: A Simple Guide to the Science of All Sciences*, Oneworld, Oxford.

Kauffman, S. A. 2008, *Reinventing the Sacred: A New View of Science, Reason and Religion*, Basic Books, New York.

Kauffman, S. A. 1993, *The Origins of Order: Self Organization and Selection in Evolution*, Oxford University Press, New York.

Kauffman, S. A. 1995, *At Home in the Universe: The Search for Laws of Self-Organization and Complexity*, Oxford University Press, New York.

Kauffman, S. A. 2000, *Investigations*, Oxford University Press, Oxford; New York.

Khlebopros, R. G., Okhonin, V., and Fet, A. I. 2007, *Catastrophes in Nature and Society: Mathematical Modeling of Complex Systems*, World Scientific, Hackensack, NJ.

Kohler, T. A., and Gumerman, G. J. 2000, *Dynamics in Human and Primate Societies: Agent-Based Modeling of Social and Spatial Processes*, Oxford University Press, New York.

Kollman, K., Miller, J. H., and Page, S. E. 2003, *Computational Models in Political Economy*, MIT Press, Cambridge, Mass.

Krugman, P. R. 1996, *The Self-Organizing Economy*, Blackwell Publishers, Cambridge, Mass., USA.

Langton, C. G. 1995, *Artificial Life: An Overview*, MIT Press, Cambridge, Mass.

Levin, S. A. 2003, Complex adaptive systems: Exploring the known, the unknown and the unknowable, *Bulletin-American Mathematical Society*, **40**(1), pp. 3–20.

Marten, G. G. 2001, *Human Ecology: Basic Concepts for Sustainable Development*, Earthscan Publications, London; Sterling, VA.

Miller, J. H., and Page, S. E. 2007, *Complex Adaptive Systems: An Introduction to Computational Models of Social Life*, Princeton University Press, Princeton, New Jersey.

Mitchell, M. 2011, *Complexity: A Guided Tour*, Oxford UP, New York, NY.

Nicolis, G., and Prigogine, I. 1989, *Exploring Complexity: An Introduction*, W. H. Freeman, New York.

Pascale, R. T., Millemann, M., and Gioja, L. 2000, *Surfing the Edge of Chaos: The Laws of Nature and the New Laws of Business*, New York: Crown Business, c2000.

Resnick, M. 1994, *Turtles, Termites, and Traffic Jams: Explorations in Massively Parallel Microworlds*, MIT Press, Cambridge, Mass.

Robb, J. 2007, *Brave New War: The Next Stage of Terrorism and the End of Globalization*, John Wiley & Sons, Hoboken, N.J.

Sageman, M. 2004, *Understanding Terror Networks*, University of Pennsylvania Press, Philadelphia.

Sawyer, R. K. 2005, *Social Emergence: Societies as Complex Systems*, Cambridge University Press, New York.

Sornette, D. 2003, *Why Stock Markets Crash: Critical Events in Complex Financial Systems*, Princeton University Press, Princeton, N.J.

Strogatz, S. H. 2000, *Nonlinear Dynamics and Chaos: With Applications to Physics, Biology, Chemistry, and Engineering*, 1st pbk. print ed., Westview Press, Cambridge, MA.

Suleiman, R., Troitzsch, K. G., and Gilbert, G. N. 2000, *Tools and Techniques for Social Science Simulation*, Physica-Verlag, Heidelberg; New York.

Tesfatsion, L. 2003, Agent-based computational economics: modeling economies as complex adaptive systems, *Information Sciences*, **149**(4), pp. 262–268.

von Bertalanffy, L. 1969, *General System Theory; Foundations, Development, Applications*, G. Braziller, New York.

Waldrop, M. M. 1993, *Complexity: The Emerging Science at the Edge of Order and Chaos*, 1st Touchstone ed., Simon & Schuster, New York.

Part 1

Complex Adaptive Systems in the Humanities

Chapter 1

Philosophical Analysis in Modeling Polarization: Notes from Work in Progress

Patrick Grim,[a,b] Aaron Bramson,[c,d,e] Daniel J. Singer,[f] Steven Fisher,[b] Carissa Flocken,[g] and William Berger[h]

[a]*Group for Logic and Formal Semantics, Department of Philosophy, Stony Brook University, Stony Brook, NY 11794-3750, USA*
[b]*Center for Study of Complex Systems, University of Michigan, 321A West Hall, 1085 S. University Ave., Ann Arbor, MI 48109-1107, USA*
[c]*Riken Brain Science Institute, Laboratory for Symbolic Cognitive Development, 2-1 Hirosawa Wako City, Saitama 351-0198 Japan*
[d]*General Economics, Gent University, 2 Tweekerkenstraat 9000 Gent, Belgium*
[e]*Software and Information Systems, University of North Carolina Charlotte, 9201 University City Blvd, Charlotte, NC 28223 USA*
[f]*Department of Philosophy, University of Pennsylvania, 433 Cohen Hall, Philadelphia, PA 19104-6304, USA*
[g]*Center for Study of Complex Systems, University of Michigan, 321A West Hall, 1085 S. University Ave., Ann Arbor, MI 48109-1107, USA*
[h]*University of Michigan, Department of Political Science, 505 S. State St, Ann Arbor, MI 48109, USA*

patrick.grim@stonybrook.edu

In this chapter, we outline major stages in an ongoing attempt to model the dynamics of polarization. The emphasis throughout is on the role of conceptual issues in computational modeling, firmly in a long tradition of philosophical analysis. In order to model opinion polarization, it turns out, one must distinguish between, and construct measures for, nine very different senses of polarization

Complexity and the Human Experience: Modeling Complexity in the Humanities and Social Sciences
Edited by Paul A. Youngman and Mirsad Hadzikadic

ISBN 978-981-4463-26-3 (Hardcover), 978-981-4463-27-0 (eBook)
www.panstanford.com

that appear in the literature. The model we construct is used to illustrate both overlaps and distinctions between those different senses. We think this case has immediate implications for the study of polarization across various fields. We hope it may also serve as an example of the importance of conceptual analysis within computational modeling generally.

1.1 Introduction

Computational modeling and computer simulation have quickly established themselves not merely as useful add-ons, but as core tools across the range of both the physical and social sciences. We consider computational modeling to be a promising approach to a range of philosophical questions as well, and to questions that sit on the border between philosophy and other disciplines [1–5]. Questions regarding the transference of belief, social networks, and opinion polarization fall in the latter category, bridging epistemology, social philosophy, sociology, political science, network studies and complex systems. These are the focus of our current research.

Our purpose here is not to sing the praises of computational modeling as a new philosophical technique. Our purpose is rather to emphasize the continuity of computational model-building with the long philosophical tradition of conceptual analysis [6–8]. With reflections from the process of building a specific model, we want to emphasize two points: (1) the work of constructing a computational model can serve the philosophical ends of conceptual understanding, in part because (2) attempts at computational modeling often require clarification of the core concepts at issue.

Our long-range goal is an agent-based model adequate to the social dynamics evident in data on opinion polarization from social psychology to political science. Agents are initially connected in a random network and randomly placed on a spectrum of opinion modeled on the [0,1] continuum, updating their views by a weighted averaging of those they trust, close to them in opinion. Those to which he is connected who are closer to an agent's current view and thus more trusted prove more influential; those farther away less so. In this regard our model extends that of Hegselmann and Krause [9], though employing a more realistically random rather than complete network structure and using degrees of influence in place of an artificially sharp threshold. Though the final form of our model does show the emergence of certain types of polarization given certain scalings of trust, our emphasis here is on the conceptual distinctions and decisions crucial throughout the modeling process itself.

1.2 Computational Modeling and Philosophical Analysis

In their final form, papers in scientific computational modeling always look perfect: They appear to be the work of a rational investigator who thought things through step by step in advance: from methods, to results, to discussion and conclusion. It is appropriate that these papers look that way—beneficial with regard to brevity, evaluation, and use in future work. That is how we want our work on belief networks and polarization to look eventually.

Of course, the polished published form of a paper can give an entirely misleading impression of the research trajectory—the impression that both the conceptual work at issue and the path of design and programming were neat, tidy, and foreordained. Almost inevitably, they were not. We will use our current work in progress as an example. Here, unlike its future final form, we will lay out the research in something more like real time, complete with fits, starts, and second thoughts. A key point is that those fits, starts, and second thoughts often indicate the need for philosophical analysis in a fully traditional sense. Computational modeling calls for and enforces a full and explicit conceptual understanding of what it is one is trying to model. To employ computational techniques, one must have such an understanding of what it is one is trying to find out, within which parameters, with which background assumptions, and why.

We offer our current work on belief polarization as a case in point. The history of this project is one in which we have repeatedly had to ask which abstract representations of social information contact were plausible. We have had to ask and ask again whether certain modeling assumptions were realistic portrayals of belief and trust, and whether it matters the extent to which they clearly were not. The history of the project is one in which we have repeatedly had to return to questions of how to define and measure the phenomenon we were after, and even whether there was just *one* phenomenon at issue. This exploration, which is at the edge of various sciences, has repeatedly demanded far more than computational resources. Flying under the colors of updating algorithm design and definition of quantitative measures, for example, we repeatedly found ourselves doing just good old-fashioned conceptual analysis in a new-fangled computational terminology, so to speak.

1.3 Understanding Polarization: Initial Motivations

Our analysis deals with the polarization of beliefs in society. We started with the impression that the increased polarization of America was an agreed and established sociological fact. Everybody talks about it and a range of books are written about it [10–13], so we thought it must be real.

The idea was to use the tools of agent-based modeling to try to understand such polarization better. Our goal was to understand the factors that influence polarization: factors necessary for polarization and perhaps a handful of factors sufficient for polarization. In the end, the hope is to find social measures that could be used to reduce polarization.

At the beginning, we had a hunch that increased polarization in America might have something to do with the structure of media sources. The core idea was the following: We seem to have been less polarized when there was essentially one source from which everyone got their news: the evening news on ABC, NBC, and CBS. The news coverage on the three major networks was essentially interchangeable—all a version of Walter Cronkite. All followed a journalistic code that insisted that editorializing be kept strictly separate from reporting.

News is no longer like that. Fox News and MSNBC have obvious political slants, are positioned at rival ends of the political spectrum, and do not seem to care where journalism leaves off and the editorial begins. Perhaps the change in where we get our news has something to do with why America is so polarized.

Such was the initial motivating hunch. Could a model illustrate how different media configurations might influence the dynamics of popular opinion? Could it show us whether split news media was an easy route to polarization, or even a possible route? Could it give us hints as to what kinds of factors might ameliorate or reduce polarization?

Some of our earlier work involved networks of artificial agents whose beliefs were modeled as numbers between 0 and 1 and who updated those beliefs in terms of the other agents with whom they had contact. We used that abstraction in the context of investigating infection, belief transference, and genetic crossover as alternative modes of information diffusion on networks. That work saw final presentation in polished form [14–16].

In earlier work we also used a more complicated version of that kind of belief updating in building models of information networks for black and white communities, based on data in the Greater Pittsburgh Random Household Health Survey. In that model we had also used data on trust: What kind of trust do members of each community put in information they receive from the government, for example, from their friends and family, from their church or religious leaders? (Figs. 1.2 and 1.3).

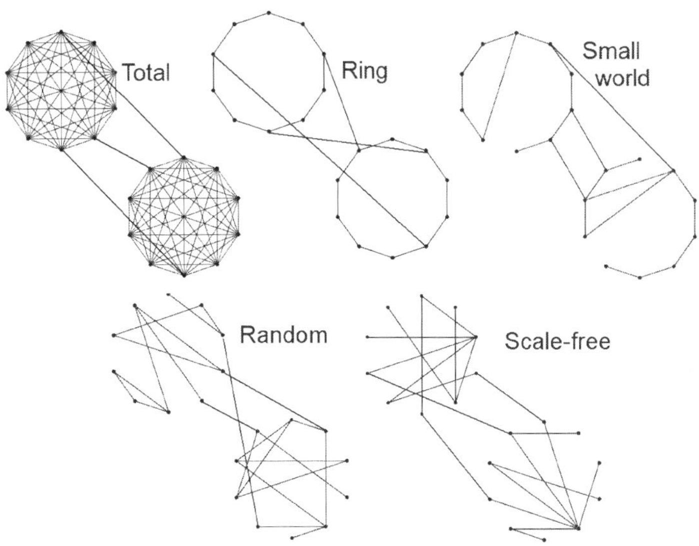

Figure 1.1 Types of linked sub-networks used in previous work on belief and infection dynamics [14–15].

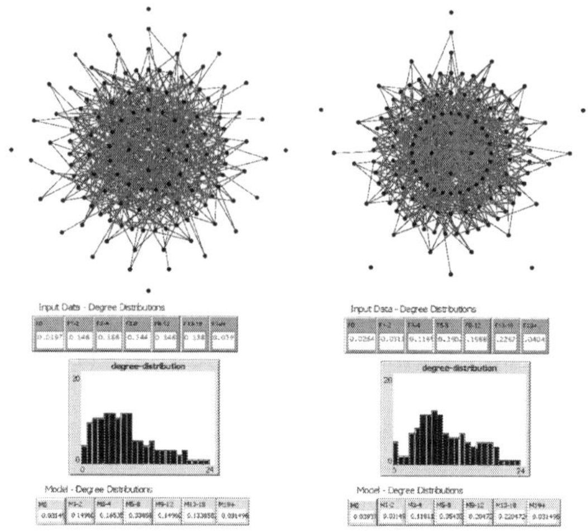

Figure 1.2 Histograms and networks constructed to match degree distributions drawn from data within the Black and White communities, Pittsburgh Random Household Health Survey [17–18]. Degrees of contact with friends and family are shown for the Black community on the left and the White community on the right.

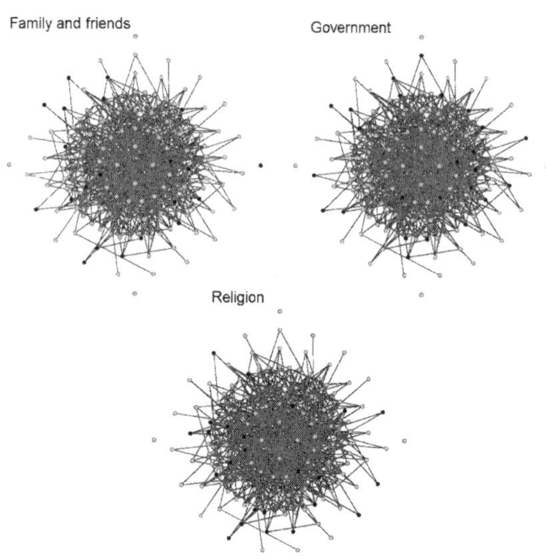

Figure 1.3 Trust levels in the Black community correlated with network position, Pittsburgh Random Household Health Survey. Black nodes indicate low trust; gray nodes indicate high trust [17–18].

This last piece of work had shown patterns of belief polarization in the two communities given conflicting input from, for example, governmental and religious sources. Why not apply the computational techniques developed in this earlier work, geared to belief change on networks and the effect of trust, in order to try to understand opinion polarization more generally?

1.4 The First Models

Our initial models were built along the following lines. Model individuals are connected via a communication network. They start with randomized "beliefs" modeled as numbers between 0 and 1. They update their beliefs based on the beliefs of their neighbors in the network. The idea is simply that we are influenced by the beliefs of those around us. If my friends all confirm my beliefs, those beliefs will be reinforced. If my contacts all seem to believe something different than I do, my beliefs can be expected to shift in that direction over time [19].

In practice we made belief updating a weighted averaging of an agent's previous belief and the beliefs of other agents with whom he had informational links in the network. That model feature is clearly artificial. But it is perhaps not implausible as a rough approximation, and has numerous precedents in the literature [20–24]. What we were after was an explanatory model; as modeling assumptions go, this representation of belief reinforcement seemed a promising start.

From the beginning, however, we also wanted to build in issues of trust. Here again, the goal was to start with a simple assumption: that widely divergent opinions can strain bonds of trust [25]. If the views expressed by a particular source are views I consider radically incorrect, wrong, or misguided, then *ceteris paribus* I can be expected to discount information from that source.

Our first models therefore had two forms of updating running in tandem: a belief updating in terms of a weighted averaging of my network contacts and a trust updating based on belief distance that is reflected in those weights. Our hypothesis was that we can more fully understand the dynamics of belief polarization in terms of the interplay between (a) belief revised in terms of trust and (b) trust revised in terms of belief.

Perhaps the fact that people discount information from contrary sources is enough to explain polarization. Perhaps a single media source—Walter Cronkite, CBS, NBC, ABC—would tend to counteract that force toward polarization. Perhaps multiple media sources—Fox and MSNBC, or the infinite number of sources one can find online to reinforce any chosen—would tend to make polarization worse.

1.5 Conceptual Questions from Computational Models

It was at this point in model development, however, that our research faced unexpected difficulties, both in terms of model results and in the conceptualization of the model itself. In the first models we built, given our initial updating assumptions for belief and trust, we kept getting convergence rather than polarization. Polarization did not seem easy to produce, even with contrasting media sources. We therefore had a wonderful model illustrating the fact that everyone is always destined to come to the same view on everything—a model that explained perfectly something that we knew did not really happen.

From another direction, and independently, we began to worry about conceptual foundations. A major issue was trust. As one member of the research group repeatedly reminded us, trust can be of various forms, from various sources. Bob has great trust in the thinking of his friend Alice. He takes Alice's views seriously and pays close attention to Alice's arguments and evidence, despite the fact that they are often in wide disagreement.

The case of Alice and Bob offers a classical philosophical counter-example. It shows, quite legitimately, that trust does not correlate with belief distance alone. We have clearly over-simplified. The question, however, was whether that over-simplification is one that can be tolerated for purposes of modeling, or perhaps an assumption that could be used *ceteris paribus*. One possibility was that we could build a model in which we tracked the effect of that factor as if it *were* the only one, drawing conclusions of the explicitly hypothetical form "were trust a matter simply of belief distance..." On the other hand, it might be that such an over-simplification is a modeling assumption that goes too far, losing track of the phenomena with which we are really concerned.

We worried that belief was single-issue and one-dimensional in our model, and that trust followed suit. Our real beliefs are multiple, and our disagreements often reflect that. I may come to trust you on one issue in one hundred, despite initial disagreement, if I have learned to trust your judgment in the other ninety-nine.

All of these are conceptual issues of a type that should be familiar to philosophers: conceptual issues regarding what belief and trust are and how they change. Here those issues arise in terms of the interpretation of a computational model: are belief and trust enough like their "representations" in the formal model to allow us to draw useful conclusions from that model, or have we sacrificed so much in the course of model simplification that we have disqualified ourselves from genuine conclusions regarding the dynamics of belief?

Goals of simplicity play a significant role in evaluating models. A model is useful only if it is simpler and easier to understand than the reality it is meant to capture, but is also useful only to the extent that it matches its target in those respects relevant to the purposes of design. Whether a model has adequately captured the relevant respects, and captured them in relevantly significant degree, is always an open question [26–28].

Even waiving those interpretational concerns in the name of model simplicity, however, we faced an issue regarding trust updating that had to be resolved in order to build the model at all. If I do discount information from those who hold views opposed to mine, precisely how much should our model discount those views? Should trust updating be modeled linearly, as shown in Fig. 1.4a, or more like in Fig. 1.4b? In the latter case, what precisely should our curve of trust-discounting look like?

In both cases, τ is the distance from an agent's belief at which there is a shift from increased trust to decreased trust. Call that trust watershed the τ-point. At this juncture we faced a number of questions. One was what the τ-point should be in our model for trust updating. Another was what field of comparison should be used for such a calculation. We might increase and decrease trust on a *local* scale, with the scope of our trust updating calibrated to each individual's immediate contacts. That would mean that our individuals discount the beliefs of those *among their network contacts* most distant from them. The alternative would be to discount on a global scale, in the sense that an individual distrusts those who would be most distant from him *across the full field of beliefs*, whether or not he has immediate contact with agents widely differing in belief.

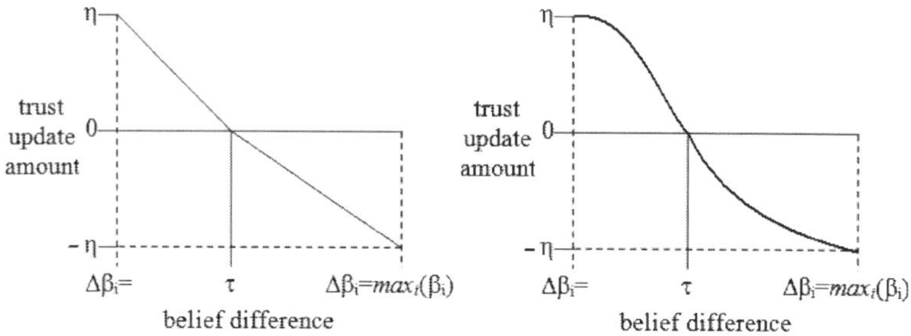

Figure 1.4 Two ways of graphing trust updating. In each case an agent increases trust as shown in an agent with a belief less than τ in distance from his own, and decreases trust as shown in an agent with a belief greater than τ from his.

1.6 Exploring the Impact of Alternatives

If we were to wait for psychologists to tell us how people update trust in terms of belief differences, whether in accord with Fig. 1.4a or 1.4b and whether against a local or global standard of comparison, we would have a long wait indeed. The truth is undoubtedly that trust updating does not occur in terms of single beliefs, is not solely in terms of belief distance, and varies in terms of update function and background comparison depending on the people and the issue involved.

That means that a predictive model of precisely what the belief dynamics will be in a particular community and a particular case is beyond us, and perhaps beyond social science generally. But prediction is not the only purpose behind computational modeling, and perhaps not the primary purpose. Explanation of general phenomena through an understanding of general mechanisms is of value even where point prediction is possible—and may indeed tell us that there will be many cases in which point prediction is *not* possible. Understanding potential dynamics in a range of cases can be as important, or even more important, than offering a specific prediction in a particular case. Understanding what factors can be expected to carry particular weight, individually or in combination, can be as important as any specific prediction based on a specific set of values for those factors.

As modelers, therefore, an alternative course of action is entirely appropriate. Our goal need not be to build some single set of realistic psychological assumptions into some specifically predictive model. What psychological assumptions are realistic may vary from person to person, from belief topic to belief topic, from community to community, and from case to case. In the attempt to understand belief dynamics in general, it is entirely appropriate to ask what the impact of alternative assumptions regarding trust will be for belief dynamics across a community and for belief polarization, for example. In that case, we are not attempting to peg the "right" value of potential factors for any particular case. We are attempting rather to figure out the relative importance of those potential factors across a range of cases, real, hypothetical, and counterfactual.

For purposes of point prediction, the level of abstraction at which we are building computational models would be a detriment; the variations in variables we are considering would simply represent a confession of ignorance. For purposes of a more general understanding of a phenomenon, the level of abstraction of models like ours can be a positive gain. With the abstract unreality of distance from the specifics that would be required for prediction in a specific case comes the power of generality. Aspects of dynamics observable in a wide range of general abstract models will be good candidates for aspects of dynamics that will hold across not just one but a range of specifiable cases. We can come to know where results change with changes in our variables.

Without being able to answer some of the questions our initial models raised, we began to make models with which we could explore what happened on some of the various options available. In some of the models we were building at this stage, polarization still refused to appear. But the scale on which trust updating was applied—the scale on which beliefs were discounted—did seem to make an important difference.

Figure 1.5 shows a typical evolution of beliefs in a network that starts with a random connection between agents of different beliefs and in which trust in other agents is discounted in terms of belief distance on a global scale. This is the evolution of beliefs in a community in which agents discount those far from their own beliefs, but far from their own beliefs in terms of the entire spectrum of opinion in the community. The result is convergence.

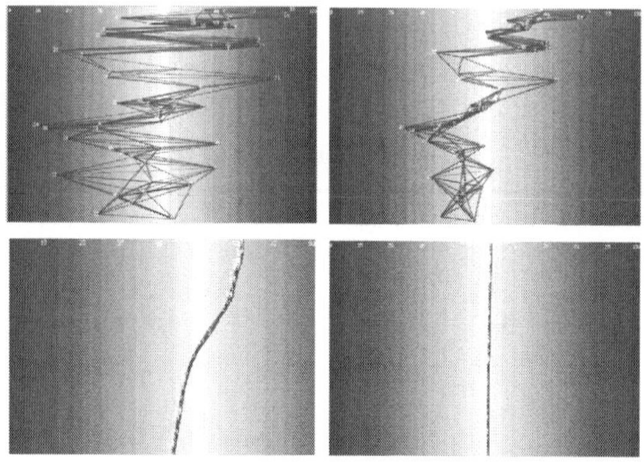

Figure 1.5 Horizontal location represents belief. Snapshots show a typical evolution of random network with global trust updating. Generations 5, 15, 25, and 30 shown.

Figure 1.6, in contrast, shows a typical evolution of beliefs in a similar random network but in which trust is discounted in terms of belief distance on a local scale. This is the evolution of beliefs in a community in which agents discount those far from their own beliefs in their own network of immediate contacts. The result starts to look more like polarization, with two distinct vertical bands in the opinion space rather than convergence on just one.

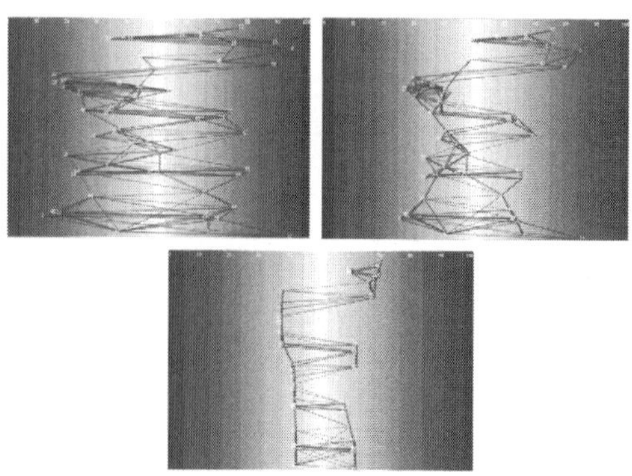

Figure 1.6 Horizontal location represents belief. Snapshots show a typical evolution of random network with local trust updating. Generations 5, 15, and 30 shown.

1.7 Philosophical Analysis in Computational Modeling: The Case of Polarization

At this point, we have the essentials of a more promising model. With networks of agents, belief updating by weighted averaging, and a range of possibilities for trust updating, we can start to measure various factors and their influence on polarization. Our goal is to determine what difference the type of network at issue makes to polarization—the relative effect of a random network of connections, for example, as opposed to a scale-free network more like many real social networks. We also want to know what difference the shape of trust-updating makes to polarization? We are currently working with the linear graph because it is easier to handle. But even given that shape, we want to know what difference a shift in τ makes: What the effect on polarization is of discounting beliefs at a distance of 0.5 from an agent's current view as opposed to 0.4 distant or 0.3 distant.

The exploration of parameters such as these forms the core of our work in progress. That work is currently qualitative, eyeballing the belief distributions that those parameter differences make, just as we invited the reader to eyeball them in the figures above.

What we would like in the end, however, is something more: a quantitative take on questions of belief dynamics, network structure, media effects, and the issue at hand. Within a range of abstract model assumptions, we would like to know just how much each of these factors can be seen to contribute to polarization. For that, we need a quantitative measure of polarization. But another conceptual difficulty arose in that regard.

As indicated in the introduction, we started with the impression that the increasing polarization of America was an established sociological fact. Everybody talks about it, a range of books are written about it, so it must be real, we thought. *Has* polarization in America increased? In order to answer such a question, it is important to know exactly what people mean when they talk of polarization. It is important to know whether there is just one thing they mean, or whether there are various senses of the term. Given different types of polarization or senses of the term, it will be important to know how one is to measure them. If you try to build a model, however simple, in which you measure polarization, such abstract conceptual questions become immediate and pressing.

A major task we have faced is simply teasing out different senses of "polarization" that appear at various points in the sociology and political science literature, but which clearly are not distinguished. Often entire articles appear on the topic of polarization, but with little attempt to make it clear what precisely is meant by the term. A real understanding of the phenomena at issue demands that we do better. The methodology of computational modeling strengthens that demand. Without claim to completeness, the following is a brief catalog of senses of the term in the literature that we have found it necessary to distinguish, and which we intend to pursue in quantitative form in further modeling:

Polarization type 1: Spread

Polarization is measured in terms of the range of opinions. One might therefore ask: How far apart are the extremes? In one of the best sociological pieces on the issue, DiMaggio, Evans, and Bryson [29] call this "dispersion": "The event that opinions are diverse, 'far apart' in content." The authors also outline a dispersion principle: "Other things being equal, the more dispersed opinion becomes, the more difficult it will be for the political system to establish and maintain centrist political consensus" ([29], p. 694).

In our model, we can measure polarization in the sense of spread as the belief level of the agent with the highest belief value minus the belief level of the agent with the lowest belief value. Polarization in this sense, however, does not consider whether the agents with minimum and maximum beliefs are extreme case outliers or the edges of large clusters. Spread is also independent of any measure in terms of groups; even if the minimum and maximum agents are representative of groups at the ends, the measure will ignore any groups in between. Although polarization in the sense of spread is important, it is also clear that we will want to measure other aspects of the phenomenon as well.

Polarization type 2: Distinctness

If we can identify different belief or attitude groups—clusters along a scale, for example— how distinct are these factions? Unlike polarization in the sense of spread, polarization in the sense of distinctness is a measure explicitly defined in terms of groups. What matters here is how clearly distinct those groups are, regardless of the distance between them. DiMaggio and his co-authors call this "bimodality" People are polarized in this second sense "insofar as people with different positions on an issue cluster into separate camps, with locations between the two modal positions sparsely occupied" ([29], p. 694).

One way to measure distinctness would be to rank the groups in order of their mean belief values and then perform pair-wise comparisons of the distributions using the Kolgomorov–Smirnov (KS) two-sample test [30–31]. This non-parametric method examines two sets of data and determines the probability that they were drawn from the same distribution, without making any assumptions about what those distributions might be. The resulting *p*-values for their being separate distributions act as measures for how distinct the groups' beliefs are. A related *N*-sample test or Bayesian method can extend that approach for any number of groups.

There is no necessary connection between polarization in sense 1 and 2; between spread and distinctness. A population might have a very diverse set of views on an issue without particular clusters emerging around any particular view. But there is no necessary disconnection, either. Attitudes toward abortion between 1970 and 1990 show both a great spread and distinctness, for instance. In their words, "If attitude polarization entails increased variance, increased bimodality, and increased opinion constraint, then only attitudes towards abortion [among those considered in the article] have come more polarized in the past twenty years, both in the public at large and within most subgroups" ([29], p. 738). "No issue represents contemporary social conflict as vividly as does abortion, the struggle over which has become symbolic of the so-called culture wars ..." ([29], p. 715).

In other sociological work, Bartels argues that voting behavior shows increased distinctness between political groups since the 1950s. Bartels demonstrates that party identification increased sharply in the 1990s, with both strong and weak identifiers increasing along with a corresponding down-tick in the number of voters that identify as independents ([32], pp. 36–37). The trend identifies a growing distinctness of the political parties along with the diminishment of independent, non-affiliated voters in the middle. The impact of distinctness on presidential and congressional races has been greater than at any time since the mid-sixties ([32], p. 42).

Polarization type 3: Uniformity within groups

How diverse are opinions *within* each group? In contrast to distinctness, this measure looks at uniformity within, rather than between, groups. The more single-minded or unanimous views are within distinct groups, the greater this sense of polarization between them. A suggestive measure is absolute deviation. The smaller the variance within distinct groups, the greater the sense of polarization across the population.

Increased uniformity as a measure of polarization is clear in the Congressional voting records of the major parties. Between 1969 and 1976—the Nixon and Ford years—the rate at which Republicans voted along party lines was about 65% in both the House and the Senate. The same was true of Democrats. Between 2001 and 2004, under George W. Bush, Republicans voted with their party 90% of the time. Democrats voted with their party 85% of the time [10].

Baldassarri and Gelman [33] also find increasing party polarization. They write, "Looking separately at trends among Republican and Democratic voters... we find clear evidence of increasing constraint within issue domains, especially among Republicans. In fact, Republicans have become more consistent on economic and civil rights issues, while Democrats have lost constraint on these issues and become a bit more coherent in their moral views. In both groups of voters, the constraint is growing faster than in the populace as a whole" ([33], p. 436). On numerous accounts, the Democratic and Republican parties have become more internally uniform.

Polarization type 4: Size disparity

A society that has one dominant opinion group with a few small minority outliers seems less polarized than one with a small number of comparably sized competing groups. Groups are more polarized in this sense if the different beliefs are held by equal numbers of people. Using the notation that G is the set of groups, and γ_i is the size of group i, size disparities can be measured by calculating the absolute deviation: $1/(2N) \times \Sigma|\gamma_i - \mu_G|$. What this formula gives us is the normalized sum of distances from the mean community size; it equals zero when all the groups are the same size and increases the more groups differ from the mean size. It maxes out at 1 as the number of groups and size differences go to infinity, making it a nice measure for comparison across different configurations.

Views on women's role in public life are no longer as polarized in this sense as they once were, even though there are small groups who continue hold anti-feminist views that were once much more common. In the past, major portions of the population once fought racial integration vociferously. Even if the views represented there are still held by some, polarization on the issue of racial integration has clearly decreased.

Polarization type 5: Coverage

We think of polarized societies as having a few tightly packed sets of beliefs. The inverse of this, a broad spectrum of beliefs, can be captured in a variety of ways. One example is the proportion of the belief spectrum held by members of society. The larger the areas of unoccupied belief space, the more polarized the society. The more focused and less diverse the beliefs in a society are, the more polarized it is.

A simple way to envisage the measure in a discrete instantiation is to think of the spectrum of possible beliefs between 0 and 1 as divided into small bins of size d (e.g., d = 0.01 or normalized by setting d to 1/the number of agents). We can then measure coverage in terms of the proportion of bins filled. Alternatively, we might want a continuous measure over the belief space. This can be done by summing the amount covered by d-diameter halos around each agent; i.e., any portion of the belief space that is within d of an agent is considered covered; the rest is uncovered. Polarization in the sense of coverage is related to dispersion, but does not include the shape of the belief dispersion. We might therefore think of coverage as a sub-measure of global dispersion, measuring how much dispersion there is without measuring its location.

Polarization type 6: Regionalization

While polarization in the sense of coverage represents how much belief dispersion there is without accounting for where beliefs are dispersed, we might also want to measure certain aspects of belief regionalization without attending to the belief area covered over all. In considering small bins of possible belief, for example, we might mean by polarization not how few bins are filled, but the extent to which there are regions of empty bins *between* regions of bins that *are* occupied.

With 100 bins, for example, there might be three different cases: (a) that in which bins 0–50 are the only bins filled, (b) the situation in which bins 0–25 and 30–55 are filled, and (c) the situation in which 5-bin regions are filled, separated by 5-bin holes: regions 0–5, 10–15, 20–25, 30–35, ... are the only ones filled. Each of these will be equally polarized in the sense of polarization as coverage. Counting the number of empty regions between regions of occupied spaces, however, gives us a measure of polarization in which (c) is more polarized than (b), which is in turn more polarized than (a). Regionalization seems a further intuitive sense of polarization well worth quantifying.

It should be noted that regionalization per se does not distinguish between the case in which (b) bins 0–25 and 30–55 are filled, and (d) that in which 0–25 and 75–100 are filled. In terms of regionalization that may be exactly what we want: beliefs in the two cases are regionalized in precisely the same sense, though the groups are farther apart in the sense of spread. Senses 1 through 6 of polarization can all be seen in terms of histograms of beliefs on a single issue across a population. But there are other senses of the term that are multiple-opinion or network-based.

Polarization type 7: Multiple opinion convergence

Given polarized groups on issue A, are these same groups polarized on B, C, and D? The more interlocked rival beliefs are within rival groups, the greater the polarization across the community. Fiorina and Abrams [34] note that intra-group polarization in this sense may increase even though population distributions on particular issues may not change. Bishop [35] notes that individuals may move to "neighborhoods where others have similar political views, changing their partisan identifications to match their ideological and issue positions" ([35], p. 578).

Polarization type 8: Community fracturing

Sub-communities may be polarized simply in the sense that there is little or no communication between them. Even if two separated communities have identical and uniform beliefs, that uniformity may be coincidental and temporary. Varshney [36] demonstrates how group interactions ameliorate levels of inter-group violence, and conversely, how group isolation increases the likelihood of violence. Varshney's central claim is that "pre-existing local networks of civic engagement between two communities stand out as the single most important proximate cause" for the difference between peace and violence ([36], p. 9). Put another way, cities with social networks that connected Hindus and Muslims through the same institutions were much less likely to see outbreaks of ethnic violence than cities in which Hindus and Muslims belonged to distinct civic institutions.

1.8 First Results and Work in Progress

We think we have made some advance, along the lines above, in the conceptual foundations important to model building with an eye to understanding polarization. Simple assumptions of a single belief scale and belief updating will remain, but with a range of variability to be explored in (a) trust updating functions with (b) different τ values against (c) local and global scales, with further variations in (d) social network structures and sizes, (e) initial configurations, and (f) media sources and effects.

Our measures in exploring variations in those parameters will be measures of polarization in the distinct conceptual senses outlined above. Here we offer a sample of the kinds of results at issue. We begin with a random network of 50 agents each of which is initially assigned a belief between 0 and 1. We use a simple linear function for belief updating. That function is "tune-able": it may be when a contact is within 0.2 of an agent's belief that his trust in that contact increases, and beyond 0.2 that he begins to discount input from that source. Or that τ-point may be wider: It may be a distance of 0.3 that marks the difference, or any other number.

Consider now two variations. In one, the τ-point is marked on a scale calibrated to the entire spread of beliefs across the population. In that case the belief spread of my particular contacts may not be as important. Relative to the range of opinions across the population, all of my friends may think pretty much like me. We will have a mutual opinion admiration society, increasing trust in each other and influence on each other based on trust. This first variation is a "global" updating model. I tend to trust individuals with beliefs like mine, gauged against the whole spread of public opinion.

Consider a second variation that differs only in the scale on which trust updating is measured. In this case τ-points 0.2, 0.3, and 0.4 are not measures across the whole spread of beliefs within the population at large. They are measures across only the spread of beliefs of my immediate contacts. In this case it will be guaranteed that one of my contacts is the farthest out—and I will decrease trust in that individual no matter how close our beliefs on the "objective" scale of the entire spread within the population. The result is a "local" updating model as outlined above. I trust those among my contacts with beliefs like mine, gauged against the field of opinion among those with whom I am in contact.

Given the other particulars of the model—a random network of 50 agents and a linear updating function—the difference between global and local scaling makes a major difference in the emergence of polarization, in several senses. Figure 1.7 shows a sample of what happens with a τ point of 0.25 and global updating. Figure 1.8 shows by contrast what happens with a τ point of 0.25 and local updating. More complete animations for each are available at www.pgrim.org/workinprogress.

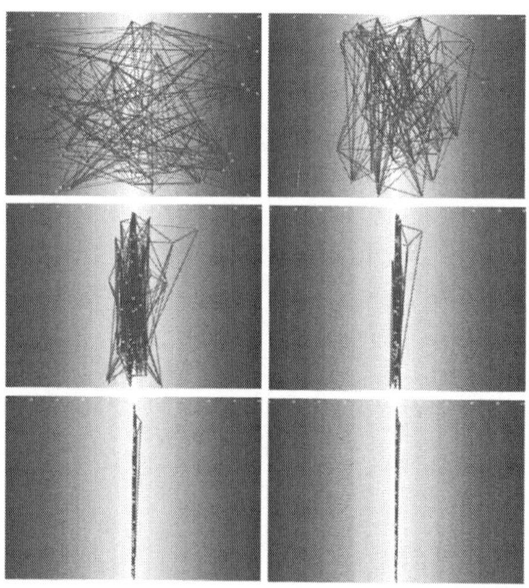

Figure 1.7 Horizontal location represents belief. Representative slides from evolution of a random array with a τ point of 0.25 and global updating. Agents update trust positively in those closest to their beliefs, update trust negatively in those farthest away, with a transition point from positive to negative update at $\tau = 0.25$. See also www.pgrim.org/workinprogress.

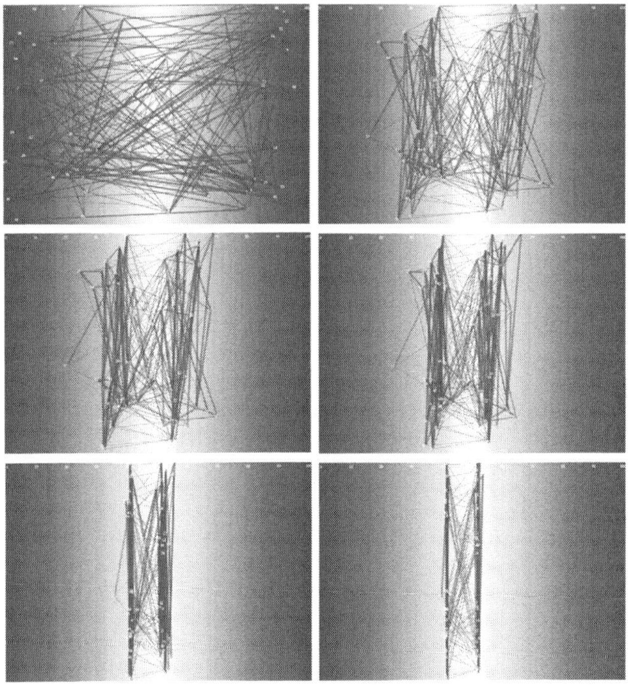

Figure 1.8 Horizontal location represents belief. Representative slides from evolution of a random array with a τ point of 0.25 and local updating. Agents update trust positively in those closest to their beliefs, update trust negatively in those farthest away, with a transition point from positive to negative update at $\tau = 0.25$. See also www.pgrim.org/workinprogress.

Figure 1.9 shows results side by side for different τ points from 0.05 to 0.75 with the same initial random seed, so that the initial beliefs in the community are the same. On the left are results for global updating. On the right are results for local updating. Global updating, it turns out, goes to belief convergence with even a very small τ value. Local updating produces polarization all the way up to a τ value of 0.5.

What these initial results indicate is that in looking for factors that favor polarization, local versus global updating can play a major role. Note also that we can distinguish many different types of polarization mentioned above in these images. In the image for local updating with a τ of 0.5, polarization is high in a number of senses. We have two major groups and a smaller intermediate group that are clearly distinct—polarization sense 2. They vary in how sharply peaked they are—polarization sense 3. The two major units are fairly equal in size, at least in this run—polarization sense 4. If network links are broken when trust falls below a certain level, it is a good guess that the networks at issue are fractured in polarization sense 6.

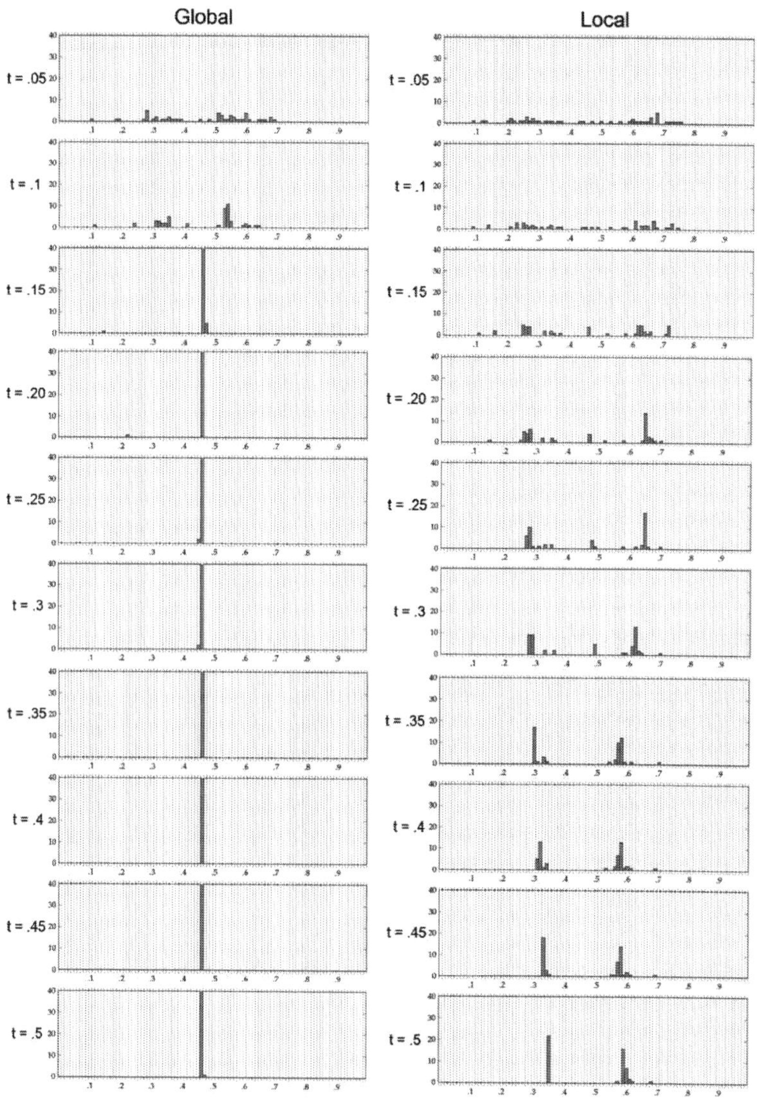

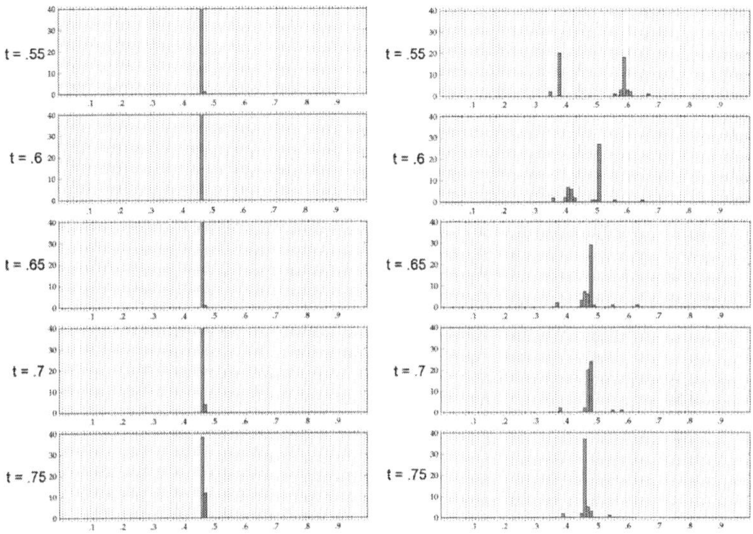

Figure 1.9 Results for global (left) and local (right) scaling with the same trust updating function (as in Fig. 1.4a) and different τ points from 0.05 to 0.75, using the same initial random seed throughout.

It is worth emphasizing that those senses of polarization are conceptually distinct. There is nothing that says logically or conceptually that polarization in one sense need accompany polarization in others. As the work progresses, it will be interesting to see whether some of these senses nonetheless appear together in modeled network dynamics much as they often seem to go together in the social dynamics that are our ultimate target.

Note also how patterns of polarization change in trust updating on a local scaling with increases in the τ point. Consider for example the patterns of polarization with a τ points at 0.2, at 0.3, at 0.4, and so on. Several senses of polarization stay the same at those points. Distinctness does—polarization sense 2. Sharpness of peak on each side stays about the same—polarization sense 3. The major units remain comparative in size—polarization sense 4. The sense of polarization that changes with increasing τ is polarization sense 1—the distance of the extremes. With increasing τ points the objective position of the two groups comes closer together. Polarization in sense 1 *slowly* decreases. In the other senses it remains fairly uniform, without decrease, until the two groups actually meet. Polarization in the other senses disappears in this progression only when polarization in sense 1 does, and only because polarization in sense 1 does.

1.9 Conclusion

All the work offered here is work in progress. Our preliminary results indicate that polarization in all the senses outlined is a complex phenomenon, sensitive to initial conditions. Global trust updating uniformly gives us consensus. Local updating clearly does not, but the clarity, extent, and patterns of polarization differ widely across runs. In a random network of 50 agents, with a linear trust update, local and global scaling marks a major difference. But other factors are of importance as well. We know that the shift from a random to a scale-free network gives a different picture—one in which that difference between local and global scaling is not so pronounced. Even population size will be important. It is a better appreciation for the role of different factors in the network dynamics, not of polarization, but of polarizations that is the wider area we want to explore.

What we have tried to indicate here is that an exploration of this form, though computationally instantiated, remains in large part conceptual in the sense that philosophical analysis has always been conceptual. We want our final results to be scientifically grounded. We hope they may offer some genuine social understanding. But in order to fill those goals they must also be philosophically sound, with a clear conceptual base.

We have also tried to make it clear that exploration of this kind often involves demands and openness to and opportunistic exploitation of the unexpected. We encounter conceptual problems we did not anticipate, which force us to distinctions and tools we did not have in advance, which lead us to build different models than we initially envisaged, which promise unanticipated results. We hope those results will tell us something genuinely new about the real social polarization we want to understand.

In the end, of course, when this is more than work in progress, we will write up our results in standard scientific fashion. We will make it look like we knew what we were doing all along, step by step, using a well-motivated methodology from a clear initial plan that produces a compelling compilation of results toward a tidy conclusion. In that final report, the crucial role of philosophical analysis in computational modeling may also go unmentioned.

Acknowledgments

This piece first appeared in the *Newsletter on Computers and Philosophy* of the American Philosophical Association. We are grateful for comments on an earlier version presented at the Human Complexity 2012 conference at the University of North Carolina, Charlotte. That conference grew in turn from an NEH Institute for Advanced Topics in the Digital Humanities: Computer Simulations in the Humanities, hosted at the University of North Carolina in the summer of 2011. Research supported in part under a MIDAS grant NIH 1U54GM088491-01, "Computational Models of Infections Disease Threats," administered through the Graduate School of Public Health at the University of Pittsburgh.

References

1. Burkholder, L. (1992). *Philosophy and the Computer* (Westview Press, Oxford).
2. Bynum, T., and Moor, J. (Eds.) (1998). *The Digital Phoenix: How Computers Are Changing Philosophy* (Blackwell, Oxford).
3. Holyoak, K., and Thagard, P. (1997). The analogical mind, *Am. Psychol.*, **52**, 35–44.
4. Grim, P., Mar, G., and St. Denis, P. (1998). *The Philosophical Computer: Explorations in Philosophical Computer Modeling* (MIT Press, Cambridge, MA).
5. Grim, P. (2004). Computational modeling as a philosophical methodology. In *Blackwell Guide to Philosophy of Information and Computing* (ed. Luciano Floridi), Blackwell, Oxford, pp. 337–349.
6. Hanna, R. (2000). Conceptual analysis, *Concise Routledge Encyclopedia of Philosophy* (Routledge, London), p. 106.
7. Sandin, P. (2006). Has psychology debunked conceptual analysis? *Metaphilosophy*, **37**, 26–33.
8. Beaney, M. (2009). Analysis, *Stanford Encyclopedia of Philosophy.* http://plato.stanford.edu/entries/analysis/.

9. Hegselmann, R., and Krause, U. (2005). Opinion dynamics and bounded confidence: Models, analysis, and simulations. *Journal of Artificial Societies and Social Simulation* **5**(3). http://www.soc.surrey.ac.uk/JASSS/5/3/2.html.
10. McCarthy, N., Poole, K. T., and Rosenthal, H. (2006). *Polarized America: The Dance of Ideology and Unequal Riche* (MIT Press, Cambridge MA).
11. Brownstein, R. (2007). *The Second Civil War: How Extreme Partisanship has Paralyzed Washington and Polarized America* (Penguin, New York).
12. Hetherington, M., and Weiler, J. (2009). *Authoritarianism and Polarization in American Politics* (Cambridge University Press, New York).
13. Fiorina, M. P., and Abrams, S. J. (2008), Political polarization in the American public. *Ann. Rev. Political Sci.*, **11**, 563–588.
14. Grim, P., Reade, C., Singer, D. J., Fisher, S., and Majewicz, S. (2010). What you believe travels differently: information and infection dynamics across sub-networks, *Connections*, **30**, 50–63.
15. Grim, P., Reade, C., Singer, D. J., Fisher, S., and Majewicz, S. (2011). Information dynamics across linked sub-networks: germs, genes, and memes, *Proceedings, AAAI Fall Symposium on Complex Adaptive Systems: Energy, Information and Intelligence*, FS-11-03, AAAI Press.
16. Grim, P., Singer, D. J., Reade, C., and Fisher, S. (2012). Germs, genes, and memes: function and fitness dynamics on information networks, *ALife 13 Proceedings* (MIT Press, Cambridge MA).
17. Grim, P., Thomas, S. B., Fisher, S., Reade, C., Singer, D. J., Garza, M. A., Fryer, C. S., and Chatman, J. (2012a). Belief polarization within the black and white communities: information dynamics in data-based networks, MIDAS scientific presentations, MIDAS Network meeting, Boston, February 2012.
18. Grim, P., Thomas, S. B., Fisher, S., Reade, C., Singer, D. J., Garza, M. A., Fryer, C. S., and Chatman, J. (2012b). Polarization and belief dynamics in the black and white communities: an agent-based network model from the data. *ALife 13 Proceedings* (MIT Press, Cambridge MA), pp. 186–193.
19. Visser, P. S., and Cooper, J. (2003). Attitude Change. In *Sage Handbook of Social Psychology* (ed. Hogg, M., and Cooper, J.) pp. 211–231.
20. French, J. (1956). A formal theory of social power, *Psychol. Rev.*, **63**, 191–194.
21. Harary, F. (1959). A criterion for unanimity in French's theory of social power. In *Studies in Social Power* (ed. Cartwright, D.), Institute for Social Research, Ann Arbor, pp. 168–182.

22. DeGroot, M. H. (1974). Reaching a consensus, *J. Amer. Statistical Assoc.*, **69**, 118–121.

23. Golub, B., and Jackson, M. O. (forthcoming). How homophily affects learning and diffusion in networks. http://arxiv.org/pdf/0811.4013v2.pdf.

24. Golub, B., and Jackson, M. O. (2010). Naïve learning in social networks: convergence, influence, and the wisdom of crowds, *Am. Econ. J. Microeconomics*, **2**, 112–149.

25. Lord, C. G., Ross, L., and Lepper, M. R. (1979). Biased assimilation and attitude polarization: the effects of prior theories on subsequently considered evidence, *J. Pers. Soc. Psychol.*, **37**, 2098–2109.

26. Miller, J. H., and Page, S. E. (2007). *Complex Adaptive Systems: An Introduction to Computational Models of Social Life* (Princeton University Press, Princeton NJ).

27. Grim, P., Rosenberger, R., Anderson, B., Rosenfeld, A., and Eason, R. E. (2013). How simulations fail, *Synthese*, **109**, 2367–2390.

28. Rescher, N. (2011, 2012). How modeling can go wrong: some cautions and caveats on the use of models, Epistemology of Modeling and Simulation, University of Pittsburgh, April 2011, and forthcoming, *Philosophy and Technology*.

29. DiMaggio, P., Evans, J., and Bryson, B. (1996). Have Americans' social attitudes become more polarized? *Am. J. Sociol.*, **102**, 690–755.

30. Kaner, H. C., Mohanty, S. G., and J. C. Lyons (1980). Critical values of the Kolmogorov-Smirnov one-sample tests, *Psychol. Bull.*, **88**, 498–501.

31. Wilcox, R. R. (1997). Some practical reasons for reconsidering the Kolmogorov-Smirnov test, *Br. J. Math. Statistical Psychol.*, **50**, 9–20.

32. Bartels, L. M. (2000). Partisanship and voting behavior, 1952–1996, *Am. J. Political Sci.*, **44**, 35–50.

33. Baldassari, D., and Gelman, A. (2008). Partisans without constraint: political polarization and trends in American public opinion, *Am. J. Sociol.*, **114**, 408–446.

34. Fiorina, M. P., Abrams, S.J., and Pope, J. C. (2010). *Culture Wars? The Myth of a Polarized America* (Longmans, New York).

35. Bishop, B. (2008). *The Great Sort: Why the Clustering of Like-Minded America is Tearing Us Apart* (Houghton Mifflin, New York).

36. Varshney, A. (2002). *Ethnic Conflict and Civil Life: Hindus and Muslims in India* (Yale University Press, New Haven CT).

Chapter 2

Overcoming Moral Hypocrisy in a Virtual Society

Markus Christen

University Research Priority Program Ethics,
Zollikerstrasse 117, 8008 Zürich, Switzerland

christen@ethik.uzh.ch

Complying with moral norms increases the reputation of individuals in a society but imposes costs, or missed gains, upon the individual when confronted with temptations. Thus, moral hypocrisy—avoiding the cost of being moral while maintaining moral reputation—may be the optimal behavior of individuals, which is corroborated by psychological research and everyday experience. In this study, the effect of various social strategies—e.g., avoiding wrongdoers or disclosing hypocrites—on the prevalence of moral hypocrisy in a virtual society is evaluated. We show that diversity with respect to population and temptation density is crucial to overcome moral hypocrisy.

2.1 Introduction

Morality is a fundamental aspect of the societal organization of mankind. Standards of morality such as norms, values, and virtues

Complexity and the Human Experience: Modeling Complexity in the Humanities and Social Sciences
Edited by Paul A. Youngman and Mirsad Hadzikadic

ISBN 978-981-4463-26-3 (Hardcover), 978-981-4463-27-0 (eBook)
www.panstanford.com

pervade the practical life of humans and safeguard the society from behaviors profitable for individuals but damaging for the group. Individuals fulfilling these standards gain a high reputation. This factor is considered to be an essential component for the development of morality in hunter-gatherer societies, where each individual is strongly aware that he or she must have a positive reputation in case of future need, and painfully guards it [11]. However, moral behavior can also involve disadvantages for an individual—either direct in the sense of "missed opportunities" or indirect in the form of costs that result from punishing wrongdoers. Concerning the latter, research shows that individuals seem to be willing to pay the costs of enforcing moral norms, e.g., through altruistic punishment [7]. Furthermore, the propagation of this strategy seems to increase with size [12] and social complexity [15] of a society. However, the fact that punishment is enforced also indicates that the temptation to trespass moral norms still exists, because the violator has gains—in particular, if the violation is not detected by anyone, e.g., the case of subtle cheating [18].

Thus, a tension between moral reputation and moral action emerges: The former has a beneficial effect for the individual, as he or she becomes a respected member of the society. However, fulfilling moral standards when confronted with specific temptations imposes losses on the individual that he or she may try to avoid. This leads to moral hypocrisy: avoiding the cost of being moral while maintaining the appearance of morality. There is no reliable data on the prevalence of this behavior in different kinds of societal organization, but it can be expected that moral hypocrisy is a widespread phenomenon in modern societies where social control is less effective compared to small-scale societies [4] and where the "opportunity space" increases due to modern technologies such as the Internet [3]; dating Web sites allowing married persons to find additional sexual partners are apt examples. The potentially substantial prevalence of moral hypocrisy is corroborated by social psychology research [1,8,17] and by the everyday observation that exemplars of moral hypocrisy—e.g., if they concern sexual moral norms like adultery—attract a great deal of attention in popular media.

In this study, we analyze moral hypocrisy from a modeling point of view by focusing on social strategies—examples are avoiding or disclosing hypocrites—that are integrated in a virtual society aiming to counteract moral hypocrisy. This approach complements the current discussion on moral hypocrisy in social psychology that focuses either on the psychological factors of how individuals maintain a motivational state with the ultimate goal of appearing moral while, if possible, avoiding the costs to oneself of actually being moral [2], or why individuals' evaluations of their own moral transgressions may differ substantially from their evaluations of the same transgressions committed by others [19]. The next section presents the model and its validation. The results section illustrates the effect of various combinations of social strategies on the prevalence of moral hypocrisy for four paradigmatic scenarios. The model implements these scenarios both individually, called non-diversity condition, and as combination, called diversity condition. The concluding section contains a discussion of the relevance of the results obtained for potential real-world strategies against moral hypocrisy.

2.2 The Moral Hypocrisy Model

2.2.1 Conceptualization of Moral Hypocrisy

Our model implements the conceptual idea of moral hypocrisy by distinguishing two different types of agent-states: the reputation of the agent, either morally good (G) or bad (B), and its disposition to act toward temptations, that is, either to be tempted (T) or to resist a temptation (R). The combinations of these states offer four different behaviors to the agents: appearing good and resisting a temptation (GR; "good guys"), appearing good but being tempted (GT; "hypocrites"), appearing bad and being tempted (BT; "bad guys") and appearing bad but resisting temptations (BR; "inconsistent guys"). The model is spatial; the agents interact by comparing payoffs within their Moore neighborhood—the eight cells surrounding a central cell occupied by the agent on a two-dimensional square lattice—and follow, if allowed to do so, social strategies that may dislocate the agent on the lattice.

The payoff structure (see Table 2.1) depicts the advantage of moral hypocrisy, i.e., an agent gains most if it is tempted although its reputation is good. In other words, the model *assumes* that moral hypocrisy is the optimal behavior for a single agent within a society. Therefore, it can be expected that "hypocrites" (GT agents) dominate within this virtual society unless other factors overcome this dominance. These factors can either be generic, e.g., population or temptation density that are predefined in the beginning. Or they can be interventional, i.e., they consist of a social strategy that intends to overcome the dominance of hypocrisy.

Table 2.1 Payoff-matrix for an agent that implements one of the four behavior types of the model

		Disposition to act	
		Be tempted (T)	**Resist temptation (R)**
Reputation	Good (G)	GT (yellow): +1 for each neighbor and each temptation in Moore Neighborhood	GR (blue): +1 for each neighbor in Moore Neighborhood
	Bad (B)	BT (red): +1 for each temptation in Moore Neighborhood	BR (pink): 0

Note: The color code refers to the figures (the online version has color graphics).

The goal of this study is to assess the success of different strategies compared to a benchmark, characterized by no strategy installed, in dependence of generic factors that model social complexity through a predefined population and temptation density. Success is measured in terms of changes in the population distribution of agents that follow one of the four behaviors. The model implements a basic conceptualization of moral hypocrisy, disregarding the specific type of temptation and internal psychological mechanisms that may, for example, explain why "inconsistent" behavior could be possible.

2.2.2 Model and Social Strategies Implementation

We implemented the moral hypocrisy model in NetLogo [20] using a 41 × 41 lattice that offers space for maximally 1681 agents and an equal number of temptations. There are four model parameters: two scenario and two initialization parameters. Scenario parameters are the agent density d_a and the temptation density d_t; initialization parameters are the probability p_r that an agent has a good reputation—otherwise, it has a bad reputation—and the probability

p_t that an agent is tempted, when a temptation is present in its Moore neighborhood—otherwise, it is not tempted.

There are two versions of the model: In the non-diversity version, both agents and temptations are randomly distributed on the lattice. Also in the diversity version, agents and temptations are spread randomly, but agent and temptation densities are different in the four quadrants of the lattice such that the four main scenarios (see Section 2.3) were simultaneously present in the model.

In both versions of the moral hypocrisy model, one of the four behaviors depicted in Section 2.2.1 is assigned to each individual agents based on predefined p_r and p_t. After initializing the model in this way, each agent obtains its payoffs based on the payoff-matrix (Table 2.1) as follows: Reputation is a function of the size of the agent's community, and "good" agents obtain as many points as there are other agents in its Moore neighborhood. The same holds for temptations, i.e., an agent that is disposed to be tempted will obtain as many points as there are temptations in its Moore neighborhood. For example, a GT agent with 3 agents and 2 temptations in its neighborhood will obtain five points in an update cycle. After updating where all agents are called in a random order, each agent compares its payoff with the payoff of its neighbors. If one neighbor has a higher payoff, the agent will adopt the behavior of the "winner"; e.g., turns from GR to GT. If not, the agent keeps its behavior; in case of tie, the agent switches its behavior with probability 0.5. After updating the behavior of all agents in this way, the sizes of the populations representing the four behavior types (GR, GT, BR, and BT) are counted and the next cycle begins.[a] If the change of the GT population in a consecutive cycle is smaller than 1% of the mean of the GT populations of the 10 previous cycles, the simulation has reached a quasi-stable state and stops.[b]

To make the result independent from the specific initialization with respect to the distributions of behaviors at the starting

[a]Updating with respect to payoff calculation and following of the implemented strategy have been checked by hand in a smaller lattice (9 × 9) for up to 20 steps to ensure that the conceptualization of moral hypocrisy and the implementation of the strategies are correct.

[b]Several windows of summation for determining the quasi-stable state have been tested. It has been shown that the GT population is sufficient in order to detect quasi-stability for all populations, i.e., the population sizes do not change any more with exception of random fluctuations.

point, various initial conditions have been evaluated for each setting of population and temptation density: p_r and p_t were chosen from the interval [0.1, 0.9] in steps of 0.1, as the endpoints 0 and 1 lead to trivial solutions (see Section 2.3). This allows the calculation of the relative size of each population x_1, x_2, x_3, and x_4, for each setting of the four parameters p_r, p_t, d_p, d_t, whereas x_1 is the fraction of GR agents, x_2 is the fraction of GT agents, x_3 is the fraction of BR agents, and x_4 is the fraction of BT agents. For some investigations, we calculated the total count over all initializations of p_r and p_t for each of the four populations.

Furthermore, 11 social strategies (see Table 2.2) have been installed in the model. Those are either pure forms or combinations[c] of three basic strategies: to avoid agents tempted to seek agents with good reputation, or to disclose a hypocrite—latter means changing its behavior from GT to BT. The third basic strategy has been implemented in a local or a global form, whereas latter models the effect of mass media, i.e., everybody learns about moral hypocrisy of an agent. The "avoidance" strategy follows the intuition that people tend to avoid wrongdoers, the "seek" strategy follows the intuition that agents with good reputation are role models, and the "disclose" strategy follows the intuition that hypocrisy made public damages the reputation of the agent.

Table 2.2 Description of the social strategies (1–4: basic strategies, 5–11: combination of basic strategies) that intend to defeat moral hypocrisy

Number	Description of strategy
1	*Avoid agents that are tempted*: Every agent that has either a GT or BT neighbor moves to the closest free cell on the lattice without such neighbors
2	*Seek agents with good reputation*: Every agent that does not yet have either a GR or GT neighbor moves to the closest free cell on the lattice with at least one such neighbor

[c]In the model validation procedure, the effect of sequencing strategies has been checked for scenario B (see Section 2.3), where maximal changes in population distributions per strategy are visible. This revealed that the position of strategies 3 and 4 in the sequence do not significantly influence the population distributions (<1%), whereas there is a sequence effect for strategies 1 and 2. For the analysis, the sequence that benefits the GT population the least has been chosen (2 before 1).

3	*Disclose hypocrite (local version)*: Whenever the majority of agents in a two-degree Moore neighborhood (24 cells) of a specified GT agent is non-GT, this GT agent becomes a BT agent
4	*Disclose hypocrite (global version)*: After specifying a GT agent: whenever the majority of all other agents is non-GT, this GT agent becomes a BT agent
5	First strategy 2, then strategy 1
6	First strategy 1, then strategy 3
7	First strategy 1, then strategy 4
8	First strategy 2, then strategy 3
9	First strategy 2, then strategy 4
10	First strategy 3, then strategy 2, then strategy 1
11	First strategy 4, then strategy 2, then strategy 1

Note: For abbreviations, see Table 2.1.

These strategies do not, of course, replicate the complexity of behaviors toward wrongdoers and hypocrites in the real world. For example, the appeasing effect of public confessions of wrongdoing is not considered. Rather, they serve as an approximation for understanding the effect of various social strategies and combinations of strategies that intend to reduce moral hypocrisy. In particular, we can assess whether a strategy that seems to be preferred in modern societies—namely disclosing hypocrites by means of mass media, in particular if they are prominent figures like the former New York Governor Eliot Spitzer or the professional golfer Tiger Woods—is indeed as successful as one may assume. The strategies described in Table 2.2 have been implemented in the basic model so that each agent in each update cycle acts according to the strategy after payoff-comparison. The effect of each strategy is measured by the sum of the differences in the population distributions compared to the benchmark—the version without strategies—for all initial conditions with respect to reputation and temptation probabilities.

2.3 Paradigmatic Scenarios

During pre-tests, we have calculated the population distributions using three strategies (1, 2, and 3) for various pairs of d_a and d_t

(42 in total) and for all initial settings of p_r and p_t in steps of 0.05 including the trivial states (p_r, p_t = 0 and p_r, p_t = 1). This revealed, as expected, that in the trivial states where only one behavior type is present no changes occur, and that generic states with very low and very high p_r and p_t do not display interesting behavior. For example, for $p_t \approx 1$, we observe absolute dominance of moral hypocrisy. Using the results of the pre-tests, we established four paradigmatic scenarios for further analysis (see Fig. 2.1):

- **Scenario A—Pre-Modern:** This scenario is characterized by a low population (d_a = 0.1) and a low temptation density (d_a = 0.05). It follows the intuition that pre-modern societies consisted of small groups with high social control minimizing the number of available temptations. In the benchmark where no strategy is implemented in the model, GR dominates the overall count, whereas BT has—when the initial conditions are suitable—a fair chance to become a population of relevant size, too.
- **Scenario B—Modern Agricultural:** This scenario is characterized by a low population (d_a = 0.1) and high temptation density (d_t = 0.5). It follows the intuition that modern agriculture consists of large farms with low population density that have access to all means of modern societies in terms of mobility, communication, etc., that tend to increase the "temptation space." In this benchmark model setting, all behavior types have a chance to dominate, depending on the initial conditions.
- **Scenario C—Brave New World:** This scenario is characterized by a high population (d_a = 0.66) and low temptation density (d_t = 0.05). It implements the idea of a city-state like Singapore, for example, with a tight control regime with respect to temptations. In this benchmark model setting, GR dominates, although GT has a significant ratio in the population count.
- **Scenario D—Sin City:** This scenario is characterized by a high population (d_a = 0.66) and high temptation density (d_t = 0.5). It implements the idea of a densely populated city full of temptations. As expected, GT is by far the largest population in the benchmark setting.

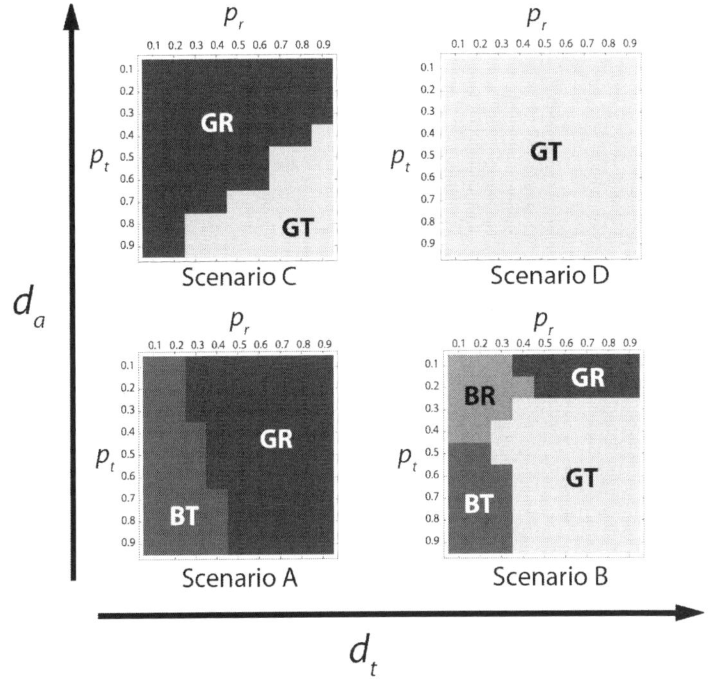

Figure 2.1 The four paradigmatic scenarios with respect to population and temptation density: Pre-modern (A), Modern Agricultural (B), Brave New World (C), and Sin City (D). For each scenario, the majority population (GR, GT, BR, and BT) after reaching quasi-stability is displayed in p_r/p_t-space according to the color code of Table 2.1. (The online version of this chapter has color graphics).

These scenarios are ideal types and only involve two of the many parameters that define human environments [9]. However, recent studies indicate that increasing population density appears to have catalysed the emergence of modern human behavior [14], making this parameter a natural choice for defining paradigmatic scenarios. And as temptation density—with respect to the number of available artifacts in different categories and behavior options, two main classes that define human environments [5]—is evidently an important factor with respect to the prevalence of moral hypocrisy, we suggest that these four ideal types of scenarios cover a wide range of possible human environments.

2.4 Defining a Measure for Population Diversity

We also analyzed the change in relative size distributions of the four populations x_1, x_2, x_3, and x_4 for specified parameter settings due to interventions by a measure for population uniformity $U(x_1, x_2, x_3, x_4)$ that is zero for $x_1 = x_2 = x_3 = x_4$ and maximal (i.e., 1) when only one population is present. The following measure fulfills these requirements:

$$U(x_1, x_2, x_3, x_4) = \frac{1}{6}\sum_{i,j=1}^{4}(x_i - x_j)^2 \tag{2.1}$$

When a strategy is compared to the benchmark with respect to population uniformity, we calculate:

$$\Delta U = \begin{cases} |U(\text{benchmark}) - U(\text{strategy})|, & \text{if no MC} \\ U(\text{benchmark}) + U(\text{strategy}), & \text{if MC} \end{cases} \tag{2.2}$$

MC stands for "majority change" that happens when for some setting of the model parameters, the population x_i is the largest in the benchmark, and $x_j \neq x_i$ is the largest after implementing the strategy. In this way, we compensate for the symmetry of U in x_i; otherwise, e.g., ΔU would be 0 when $x_j = 1$ in the benchmark and, after implementing a strategy, $x_j = 0$, whereas another population $x_j = 1$ completely dominates.

2.5 Results

We have calculated the population distributions for all four scenarios and for all 11 strategies (10 trials[d] per initial setting) both for the non-diversity and, in part, for the diversity model. In the following, we summarize the results of this analysis.

[d]Tests revealed that for 10 trials per setting of initial probabilities the standard deviation is usually in the order of less than 5% of the population size except for the case when a population has only very few members, where statistical fluctuations are naturally higher. As the populations usually are summed over all initial conditions, this effect is negligible.

2.5.1 Scenario Parameters Determine Population Distributions

The first insight provided by our model of moral hypocrisy is that the scenario parameters d_a and d_t determine the distributions of the four behavior types. In the non-diversity model, *no* strategy was able to change the majority of the dominating behavior type summed over all settings of p_r and p_t. In scenarios A and C, where d_t is low, the "good guys" (GR) dominate, whereas in scenarios B and D, where d_t is high, the "hypocrites" (GT) dominate. In scenario D, the dominance of GT is so overwhelming that no strategy was able to make any relevant impact. In the other scenarios, however, more pronounced effects are visible; exemplars of strategies that led to large changes in population distribution are displayed in Fig. 2.2. In particular, BT was able to build up a strong minority position both in scenarios A and B for selected strategies. In scenario C, GR was able to strongly increase its dominance for selected strategies.

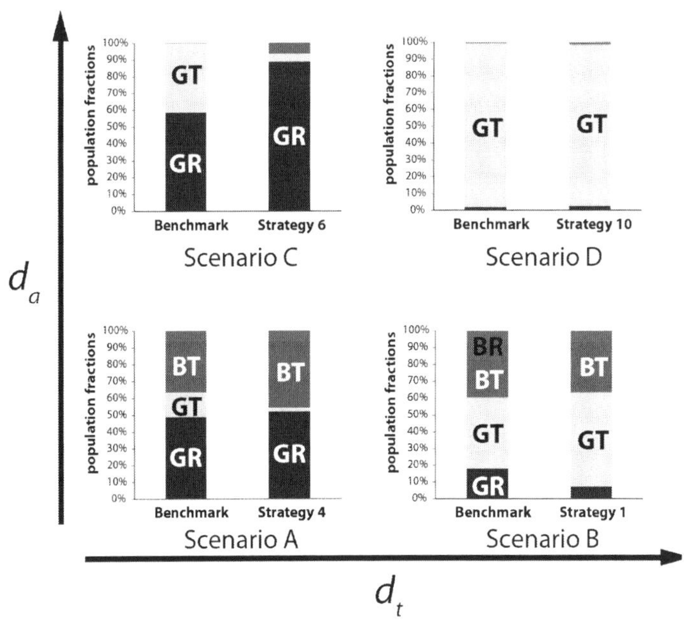

Figure 2.2 Effects of selected strategies on the population distribution in the four paradigmatic scenarios (examples): No strategy was able to change the dominance of the majority population.

2.5.2 Strategy Rankings and Conflicting Effects of Interventions

The effect of a strategy can be very different depending on the scenario in which it is implemented in the non-diversity condition. We show this by ranking the strategies according to their ability to increase the population of a specific behavior type relative to the benchmark population size (Fig. 2.3)—BR was excluded from this analysis, as this behavior type never was able to increase its population size in any strategy.

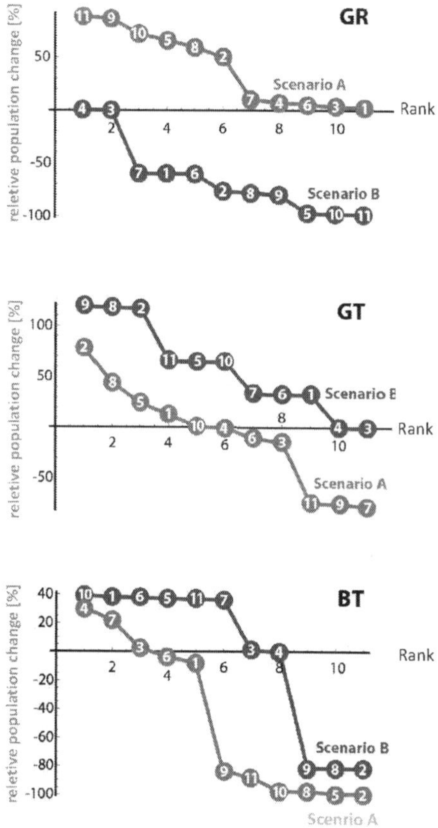

Figure 2.3 Ranking of strategies according to their effect on population size of GR, GT, and BT for scenarios A and B: Strategies optimal for scenario A are disastrous for scenario B. Numbers in dots indicate strategy according to Table 2.2.

For each ranking, we have calculated the Kendall Rank correlation—a measure of the similarity of rankings—and we display in Fig. 2.3. the two sequences with the highest dissimilarity for the "good guys" (GR), the "hypocrites" (GT) and the "bad guys" (BT). In all cases, the highest dissimilarity is obtained by comparing scenarios Pre-Modern (A) and Modern Agricultural (B). This reveals that strategies can have contradicting effects depending on the scenario. For example, strategy 11—a combination of disclosing hypocrites, seeking agents with good reputation and avoiding agents that are tempted—is optimal for GR in scenario A, but is disastrous in scenario B. The situation is analogous for strategy 9—a combination of seeking agents with good reputation and disclosing hypocrites—for GT: optimal in B, disastrous in A. For RT, strategy 10—again a combination of disclosing hypocrites, seeking agents with good reputation, and avoiding agents that are tempted—is optimal in B and disastrous in A. This means that the success of a strategy with respect to diminish the number of "hypocrites" strongly depends on the type of society, modeled by population and temptation density.

2.5.3 Strategy Effects Attributed to Four "Moral Worlds"

To get an overview of all strategies with respect to the intervention effect in the non-diversity condition, we grouped them in four classes based on how they change the population distribution: Good world strategies increase the GR population and decrease the BT and GT population, polarized world strategies increase the GR and the BT population, bad world strategies increase the GT and BT populations, and shiny good world strategies increase the GR and GT population (Fig. 2.4). We disregarded scenario D due to the small effects of all strategies in this case.

This analysis reveals tendencies with respect to the importance of main scenarios for strategy effects: First, the Modern Agricultural scenario B creates a context that promotes bad world strategies—i.e., BT and GT can often increase their weight. Second, the Brave New World scenario C creates a context that promotes polarizing world strategies increasing both the GR and the BT population. Third, the Pre-Modern scenario A creates a context that promotes shiny good world strategies.

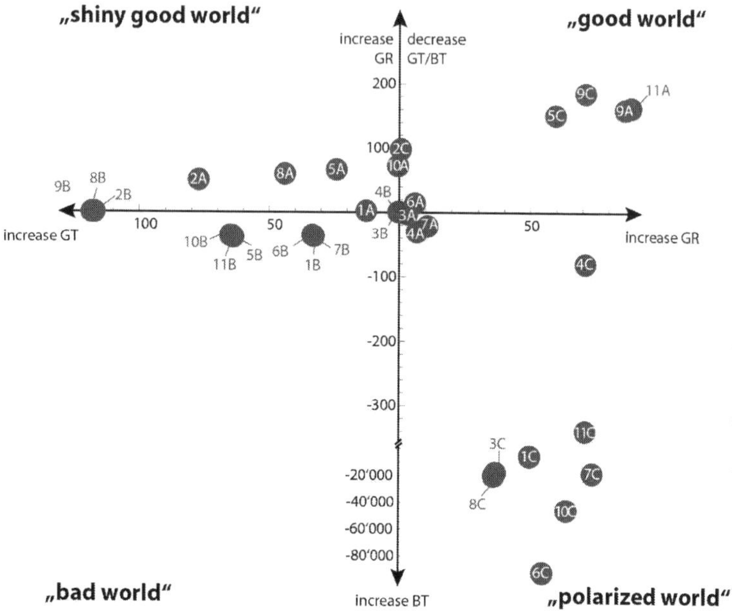

Figure 2.4 Classifying strategies in dependence of their effect in a scenario: The strong increase of the BT population in the polarized world for some strategies reflects the fact that this population is small in the benchmark. The axes display the relative population change in [%]. The axis separating "shiny good world" from "good world" has a different labeling for the right and left side.

With respect to the strategies themselves, no strategy can be attributed to only one strategy class. However, disclosing strategies in their pure form (3 and 4) tend to be polarizing, i.e., form strong minorities of BT agents. Strategy 2—seeking agents with good reputation—tends to be "shiny," which is plausible as the GT population profits from a strategy that benefits good reputation.

2.5.4 Scenario-Diversity Overcomes Moral Hypocrisy

As d_a and d_t are the dominant factors with respect to population distributions in the quasi-stable state independent from the strategy, we analyzed the effect of strategies against moral hypocrisy on populations when the model implements *all four* basic scenarios simultaneously. This diversity condition can be understood to increase the social complexity of the model, that is depending on the strategy agents may leave regions of high temptation density.

In the benchmark condition, GT ("hypocrites") is the dominant behavior for almost all initial settings, as expected by the payoff structure (Fig. 2.5). However, contrary to the findings in the non-diversity condition, 5 out of 11 strategies are able to change population majorities from GT to GR ("good guys"). Diversity seems therefore to be a crucial factor for overcoming moral hypocrisy.

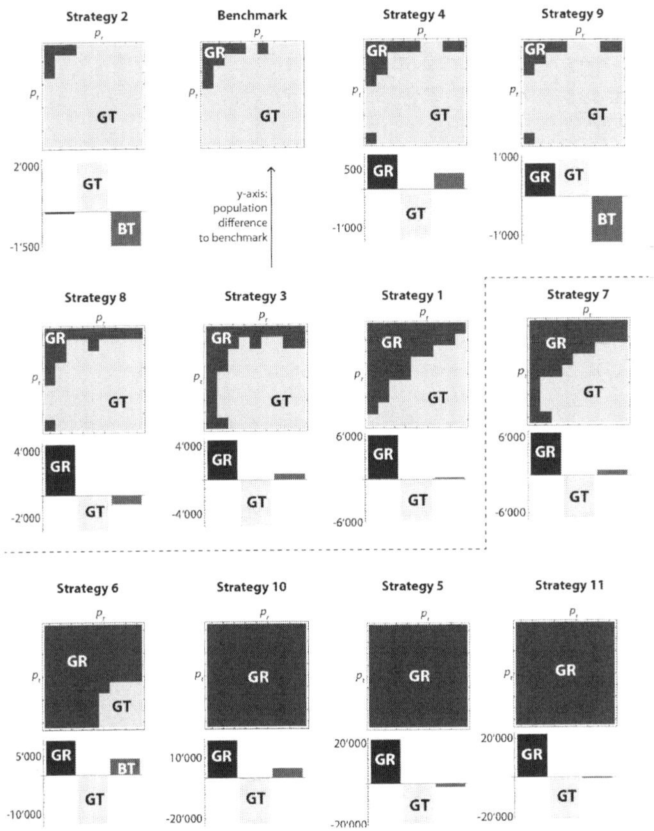

Figure 2.5 Effect of strategies in the diversity condition: Five out of 11 strategies, ranked from first row left to second row right according to increase in GR population, overcome the GT majority, indicated by dotted line. Shown are majorities in p_r/p_t-space and absolute changes in the population counts of GR, GT and BT.

It is remarkable that this majority change is caused mainly by the fact that agents move to regions where temptation density is low, although there is no built-in mechanism in the model to avoid temptations. Agents move away from other agents that are tempted, which is reflected by the fact that those strategies that include the basic strategy 1 ("avoid agents that are tempted") are successful with respect to majority change, although strategy 1 alone is not yet able to induce a majority change.

In order to further analyze majority changes in the diversity condition, we calculated the population uniformity p_r/p_t-space according to Eq. (2.1) and (2.2) for the benchmark and for all strategies. Figure 2.6 shows the mean $\Delta U(x_1, x_2, x_3, x_4)$ over all strategies compared to the benchmark. This reveals that the effect of strategies is not uniform in p_r/p_t-space. Rather, a low probability for having a good reputation p_r in the model initialization generally leads to larger population changes after strategy implementation, independent of the probability to be tempted p_t. Another local maxima in ΔU is discernible for high p_r and low p_t, i.e., an initial setting that consists mostly of GR agents ("good guys") is vulnerable for majority changes.

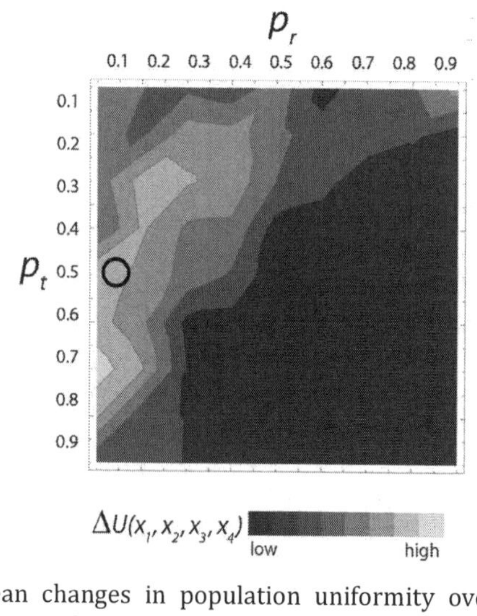

Figure 2.6 Mean changes in population uniformity over all strategies compared to the benchmark: The initial conditions affect the vulnerability of dominating populations for majority changes after strategy implementation differently. The circle indicates the parameter choice for the analysis depicted in Fig. 2.7.

Finally, we investigated the effect of population and temptation density for *fixed* p_r and p_t in the diversity condition. We changed both the total number of agents and temptations from 50 to 750 in steps of 50 by keeping the relative differences in densities in the four quadrants of the lattice. We fixed p_r and p_t such that maximal change in population uniformity can be expected—indicated by the circle in Fig. 2.6—and we chose strategy 5 as comparison to the benchmark, knowing that this strategy is able to change the benchmark majority (GT agents), see Fig. 2.5. In this way, we can analyze how population uniformity changes in d_a/d_t-space.

Figure 2.7 shows the result of this analysis. It reveals that for low d_a the "bad guys" BT are always the dominating population in the benchmark, whereas for a higher number of agents, d_t defines,

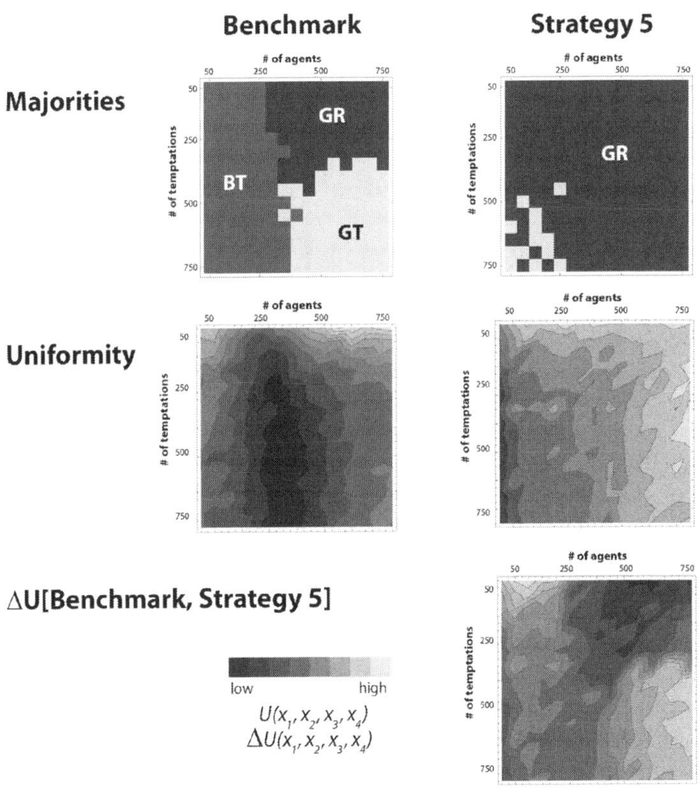

Figure 2.7 Effect of strategy 5 in the d_a/d_t-space: The change in population uniformity is highest for low agent and low temptation density as well as for high agent and high temptation density.

whether the "good guys" GR or the "hypocrites" GT dominate. Implementing strategy 5 changes, as expected, the majority count: GR becomes the dominating population. The zone with low population uniformity is shifted to the right. The highest changes in population uniformity are visible for the zone where both population and temptation density are either low or high. This means that strategy 5 is successful because it changes majorities in conditions that are usually not optimal for the "good guys" GR.

2.6 Discussion and Conclusion

This study shows that the success of strategies to overcome moral hypocrisy in a society strongly depends on the type of society in which the strategy is implemented. In the non-diversity condition, successful strategies in one scenario have disastrous consequences on the population of a specific behavior in other scenarios. Strategies thus have "side effects": For example, disclosing hypocrites in the Brave New World scenario may have the unwanted effect of also increasing the population of "bad guys." On the other hand, strategies against moral hypocrisy in a society with relatively low interaction density due to rather low population density, like the Modern Agricultural scenario, generally tend to be unsuccessful, indicating that not moral hypocrisy itself, but other aspects of social organization—e.g., interaction density—may be better targets for social interventions.

However, adding social complexity to the model changes the picture dramatically: If generic social scenarios are allowed to coexist, various strategies can overcome the dominance of moral hypocrisy. The key element for such a majority change is the strategy to "avoid wrongdoers" which has the indirect effect of moving the agents away from temptations, whereas the strategy "disclose hypocrites" does not have this effect. In this sense, our results confirm the importance of social complexity for the success of pro-social behaviors like cooperation [15,16]. Our results also accord with recent modeling studies that demonstrate the importance of agent mobility for enforcing pro-sociality [10]. It, furthermore, conforms to the philosophical insight that developing psychological competences or "virtues" to resist temptations in all circumstances may not be the best strategy to avoid unmoral behavior. Rather, a person should gain insights into the own moral psychology and—

based on this knowledge—avoid temptations for which the person is susceptible [6].

Certainly, this model will not allow finding straightforward applications for overcoming moral hypocrisy in real world societies. The primary reason for this is that the success of social strategies against moral hypocrisy depends on cultural and historical factors. Abundance of hypocrisy with respect to specific behaviors may, for example, have the effect that the moral norm itself is undermined and eventually vanishes—a phenomenon that is detectable for various norms [13]. We can also expect path-dependency with respect to the succession of several strategies against moral hypocrisy. For example, a society that first chooses to disclose hypocrites and later switches to an avoidance strategy may experience a different level of success in overcoming hypocrisy than a society that implements these strategies in reverse order. This model actually allows one to investigate path dependency of strategy implementations, for which preliminary results (not shown) indicate that this is another important factor in assessing strategies against moral hypocrisy.

We note several additional shortcomings of this approach that require further investigations: First, the current model does not take into account the inner psychological complexity of moral hypocrisy with respect to the type of temptations, for example, some agents may be tempted only by one kind of temptation. This simplification may also explain why the number of hypocrites even in a real world Sin City scenario is probably lower compared to the dominance of this population in the model. Although the model design allows in principle the integration of an inner agent psychology with respect to moral hypocrisy, it remains doubtful whether an extension of the model in this respect would indeed lead to additional insights, as no real world data is available to compare the model data of the prevalence of moral hypocrisy in a society in order to validate the refined model. Second, the model does not include the issue of "double standards" with respect to assessing the severity of moral hypocrisy, i.e., the fact that people tend to evaluate their own moral transgressions as less severe compared to the evaluations of the same transgressions committed by others. Including this aspect in the model would require a refinement with respect to agent psychology. Third, the current analysis does not involve all possible social strategies against moral hypocrisy, i.e., it is incomplete in that respect. Fourth, there are other model parameters that may become the object of further investigations. Examples are changes in the payoff-structure or non-Moorean interactions between the agents. A non-exhaustive analysis of the effect for changes in the payoff-structure by weighting either moral reputation or temptation gain higher than the other component[e] reveals, however, no qualitative change of the results, as long as the major model assumption—moral hypocrites gain the most—is fulfilled.

In summary, the analysis provides indications that the abatement of moral hypocrisy cannot rely on simple and single strategies that abstract from the generic social setting in which hypocrisy emerges. Overcoming moral hypocrisy requires context-sensitivity in order to be successful and relies on societies that are both divers and allow for social mobility.

Acknowledgments

I thank Daniel Singer for his helpful advice in developing this model of moral hypocrisy. I also thank all members of the Institute for Advanced Topics in the Digital Humanities (UNC Charlotte) for their support.

References

1. Batson, C. D., Thompson, E. R., and Chen, H. (2002). Moral hypocrisy: addressing some alternatives. *J. Pers. Soc. Psychol.*, **83(2)**, 330–339.
2. Batson, C. D., Kobrynowicz, D., Dinnerstein, J. L., Kampf, H. C., and Wilson, A. D. (1997). In a very different voice: unmasking moral hypocrisy. *J. Pers. Soc. Psychol.*, **72**, 1335–1348.
3. Benkler, Y. (2007): *The Wealth of Networks: How Social Production Transforms Markets and Freedom* (Yale University Press, New Haven).
4. Chriss, J. J. (2007). *Social Control: An introduction* (Polity Press, Cambridge).

[e]The payoff matrix has been changed as follows: emphasizing moral reputation: GT: +2, GR: +2, BT: +1; emphasizing temptation gain: GT: +2, GR: +1, BT: +2. In the benchmark condition, only the Brave New World scenario shows noticeable changes: when moral reputation is emphasized, the "good guys" profit even more, whereas when temptation gain is emphasized, the "hypocrites" gain more. This result is in accordance with the expectations.

5. Craik, K. H. (1971). The assessment of places. In *Advances in Psychological Assessment* (McReynolds, P. ed.), vol. 2, Science and Behavior Books, Palo Alto, CA, pp. 440–467.

6. Doris, J. M. (2002). *Lack of Character. Personality and Moral Behavior* (Cambridge University Press, Cambridge).

7. Fehr, E., and Gächter, S. (2002). Altruistic punishment in humans. *Nature*, **415**, 137–140.

8. Fointiat, V., Somat, A., and Grosbras, J.-M. (2011). Saying, but not doing: induced hypocrisy, trivialization, and misattribution. *Soc. Behav. Person.*, **39(4)**, 465–476.

9. Gärling, T. (1998). Introduction: Conceptualizations of human environments. *J. Environ. Psychol.*, **18**, 69–73.

10. Helbing, D., and Yu, W. (2009). The outbreak of cooperation among success-driven individuals under noisy conditions. *Proc. Natl. Acad. Sci. U. S. A.*, **106(10)**, 3680–3685.

11. Hrdy, S. B. (2009). *Mothers & Others: The Evolutionary Origins of Mutual Understanding* (Harvard University Press, Cambridge).

12. Marlowe, F. W., and Berbesque, J. C. (2008). More "altruistic" punishment in larger societies. *Proc. Biol. Sci.*, **275**, 587–590.

13. Nichols, S. (2004). *Sentimental Rules. On the Natural Foundations of Moral Judgment* (Oxford University Press, Oxford).

14. Powell, A., Shennan, S., Thomas, M. G. (2009). Late Pleistocene demography and the appearance of modern human behavior. *Science*, **324**, 1298–1301.

15. Santos, F. C., Santos, M. D., and Pacheco, J. M. (2008). Social diversity promotes the emergence of cooperation in public goods games. *Nature*, **454**, 213–216.

16. Shutters, S. T. (2011). Punishment leads to cooperative behavior in structured societies. *Evol. Comput.*, **20(2)**, 301–319.

17. Tong, E. M. W., and Yang, Z. (2011). Moral hypocrisy: of proud and grateful people. *Soc. Psychol. Person. Sci.*, **2(2)**, 159–165.

18. Trivers, R. L. (1971). The evolution of reciprocal altruism. *Quart. Rev. Biol.*, **46**, 35–57.

19. Valdesolo, P., and DeSteno, D. (2007). Moral hypocrisy: social groups and the flexibility of virtue. *Psychol. Sci.*, **18**, 689–690.

20. Wilensky, U. (1999). NetLogo. http://ccl.northwestern.edu/netlogo/. Center for Connected Learning and Computer-Based Modeling, Northwestern University. Evanston, IL.

Chapter 3

Evolving Games and the Social Contract

Rory Smead

Department of Philosophy and Religion, Northeastern University, 371 Holmes Hall, 360 Huntington Avenue, Boston, MA 02115, USA

r.smead@neu.edu

Evolutionary game theory provides a set of tools that can be used to model the evolution of social contracts. These models traditionally take the game individuals play as fixed and consider how behavior within this game may change over time. However, one of the most important aspects of the formation of social contracts is the evolution of the game itself. This chapter presents a method for modeling evolving games that fits within the standard methods of evolutionary game theory and applies this approach to examples relevant to the evolution of the social contract. We discuss the scope and limits of this method and contrast this way of representing evolving games with other approaches that have been considered in the past.

3.1 Introduction

Thomas Hobbes describes the formation of social contracts as rational individuals giving up their freedoms to avoid the "nasty,

Complexity and the Human Experience: Modeling Complexity in the Humanities and Social Sciences
Edited by Paul A. Youngman and Mirsad Hadzikadic

ISBN 978-981-4463-26-3 (Hardcover), 978-981-4463-27-0 (eBook)
www.panstanford.com

brutish, and short" life in the state of nature. In the state of nature, there is no reason to trust other individuals and there is a general lack of cooperative interactions. Adopting a social contract changes individual incentives to make trust and cooperation possible.

In recent years, philosophers have begun to use game theory to study the formation of social contracts. If interactions take the form of a Prisoner's Dilemma (Fig. 3.1), rational cooperation is not possible: Defection is the optimal action regardless of what the other person does. However, if the game is altered in some way so as to change the incentives, such as introducing the possibility of reciprocity or punishment, then cooperation can be rationally sustainable.

	Cooperate	Defect
Cooperate	2, 2	0, 3
Defect	3, 0	1, 1

Figure 3.1 Prisoner's Dilemma.

In essence, one "solution" to games like the Prisoner's Dilemma is to alter the game: Introduce complexities or change the incentives so as to stabilize cooperation. One could interpret Hobbes as suggesting that the adoption of the social contract changes the character of social interaction—it changes the game being played. Along these lines, Skyrms describes one interpretation of Hobbes' "solution" of introducing considerations of punishment or reciprocity as changing a Prisoner's Dilemma into a Stag Hunt (Fig. 3.2) where cooperation is a rational equilibrium of the game [20].

	Cooperate	Defect
Cooperate	3, 3	0, 2
Defect	2, 0	2, 2

Figure 3.2 Stag Hunt.

This notion of solving cooperative problems by "changing the game" is repeated in recent work on the "structural solutions" to social dilemmas; see [13,16]. One can solve a tragedy of the commons by altering the rules of the game in some way, such as adopting a governing institution or using extra rewards and punishments to incentivize certain kinds of behavior. The traditional solutions of appointing a ruler or privatizing the commons can also be seen as changing the game.

The basic idea is interesting and open-ended: The game itself could change or evolve. However, there are many ways a game could change and it is not clear how to model a changing game. The adoption of some institution may change a Prisoner's Dilemma into a Stag Hunt, or some other game where cooperation is easier to attain. However, other shifts could happen as well. Perhaps cooperation in a Stag Hunt leads to over-exploitation of the environment in the long run, decreasing the payoff to cooperation and effectively creating a Prisoner's Dilemma and thereby destabilizing cooperation. After all, social contracts can both form and breakdown [4]. This is a gradual and endogenous shift that is a result of player behavior within the game.

The focus of this chapter is on this latter kind of change and the aim will be twofold. The first is to illustrate a method for modeling evolving games using existing tools in evolutionary game theory. The second is to apply these tools to model some simple cases relevant to the formation (or dissolution) of social contracts.

3.2 Evolution of Behavior in Games

Before presenting an account of how to model evolving games, it is first necessary to consider evolution of behavior within a fixed game. We will focus on two-player symmetric games. However, modeling evolving games will require considering three-player asymmetric games. Some general game theoretic notation will be helpful. A game (labeled g) consists of a set of n players, a set of strategies S_i for each player, and a payoff function for each player $\pi(\bullet)$, which specifies an individual's payoffs given a strategy profile (i, j) that includes both players' strategies. The set of mixed strategies will be represented as ΔS and payoffs with respect to mixed σ in ΔS will be calculated as expected utilities $u(\bullet)$. Let $\mathrm{Br}(\bullet)$ be a correspondence such that $\mathrm{Br}(\sigma_{-i}) = \sigma_i^*$, where $u(\sigma_i^*, \sigma_{-i}) \geq u(\sigma_i, \sigma_{-i})$ for all $\sigma_i \in \Delta S$. $\mathrm{Br}(\sigma)$ determines the set of best responses to σ. A strategy profile $(\sigma_i^* \ldots \sigma_n^*)$ is a *Nash equilibrium* (NE) just in case each strategy is a best response to the other.

To illustrate these ideas, consider the Stag Hunt game above. Here, both players have identical strategy sets, Cooperate and Defect, one is the row-player, the other is the column-player, and the payoff functions provide the numbers in the matrix. The best response to Cooperate is Cooperate and the best response to Defect is Defect. There are three NE: two pure-strategy NE (Cooperate, Cooperate) and (Defect, Defect), and one mixed strategy NE where both players play Cooperate with probability 2/3.

To consider evolution of behavior in games, it is often useful to think of large populations of individuals being paired at random with one another to play some game. We imagine that each individual is hardwired to behave according to one of the strategies in the game. Evolution can then operate on the types: the more successful types will increase in frequency relative to the less successful types [11,18,24].

More precisely, an infinite population of randomly mixing individuals that each play some pure strategy of game g. For a game with n strategies, a population state is expressed by the vector $x = (x_1 \ldots x_n)$, where $x_i \geq 0$ and $\Sigma x_i = 1$.

Note that a population $x \in X_n$ is just a vector that is identical to some mixed strategy $\sigma \in \Delta S$. This allows us to talk of a strategy's payoff against a population, or a population's payoff against another population, in the same way we talk of payoffs to mixed strategies. A strategy's "fitness" will be the driving force behind the evolutionary processes. The fitness of a population can be derived from the corresponding utility function. If x_i represents the proportion of strategy i in the population, Eq. (3.1) represents fitness of strategy s against population x.

$$F(s, x) = \sum_{i \in S} \pi(s, i) x_i \tag{3.1}$$

A population x will be said to be at a Nash equilibrium if and only if all the strategies represented in x are best responses to x. Other important solution concepts include evolutionarily stable strategies (ESS) and the closely related evolutionarily stable states (ES States) [2,23]. An evolutionary stable strategy is one such that if all members of the population adopt that strategy, it cannot be invaded by any other strategy [11,24].

While stability concepts are important, our focus will be on evolutionary dynamics. There is good reason to believe that a dynamical approach provides a more thorough method for analyzing evolutionary games [12]. One of the most common dynamics used in large, randomly mixing population settings are the replicator dynamics [22].

$$\frac{dx_i}{dt} = x_i[F(i,x) - F(x,x)] \tag{3.2}$$

$F(x, x)$ represents the average payoff in the population. Equation 3.2 states that the frequency of more successful types relative to the average member of the population will increase at a rate proportional to their success. There are two interpretations of the replicator dynamics: biological and cultural. In the biological interpretation, the dynamics pertain to differential reproduction, where more successful strategies lead to more offspring. In the cultural interpretation, the dynamics pertain to differential imitation where successful strategies lead to more frequent imitation of the strategy in the population.

By way of illustration, consider the Stag Hunt from the previous section. Each of the three equilibria corresponds to a rest point in a population governed by the replicator dynamics. A population where all players cooperate is an asymptotically stable point, a population where all players defect is also an asymptotically stable point and a population that consists of 2/3 cooperators is an unstable point that will lead to the state where all cooperate if a few more cooperators are introduced and a state where everyone defects if more defectors are introduced.

3.3 Games of Incomplete Information

In some settings, players are uncertain of some aspects of the underlying game they are facing. Players may not know the payoff structure, the strategies available to other players, or the beliefs of other players. These are games of incomplete information.

For example, suppose two players are involved in a two-strategy game but do not know which game they will be playing. It may be a Stag Hunt or it may be a Prisoner's Dilemma. If it is a Stag Hunt, the cooperative outcome is feasible. If it is a Prisoner's Dilemma, defection is strictly dominant. Here, game theory cannot provide a solution and cannot recommend a strategy to be played without further specification.

John Harsanyi developed the most popular approach to studying games of incomplete information [9]. Harsanyi suggested that we consider a game of incomplete information as an extended game with imperfect information. In the example above, we can imagine that there is a third player "Nature" who chooses which game is being played. Nature chooses the Stag Hunt with probability p and the Prisoner's Dilemma with probability $1-p$. The players must then choose a strategy without knowing for sure which game they are in. In more complicated games, individuals may adjust their beliefs about which state they are in if they receive some indication or evidence during the game [8]. This modified game is referred to as a "Bayesian game."

Harsanyi's basic approach, that of treating Nature as a player, will be helpful when we want to think about how games may change over time. An important point to note, however, is that Nature has a fixed "strategy" in these games and does not alter her probabilities depending on the strategies of others (or other factors). In the model presented below, we can imagine that Nature changes the underlying probabilities over time; we will treat Nature as evolving.

3.4 Evolution with Incomplete Information

It is possible to use evolutionary ideas in an analysis of Bayesian games. This may take many forms, such as using population dynamics, e.g., the replicator dynamics, to model strategic change, or implementing various learning processes that agents may be using. For example, Ely and Sandholm introduce best response dynamics into Bayesian games and study the relationship between dynamics and static equilibria concepts [7]. Zollman also looks at evolution within this kind of "complex environment" using the replicator dynamics [26]. He examines the evolution of fair-division in two different bargaining games where individuals face both games and must choose a single strategy.

One notable case of an approach that focuses on learning processes is due to Mengel [14]. Mengel proposes a model in which individuals need to apply reinforcement learning to a game, but there is not a single game being played. In this model, individuals must pay some cost to be more finely discriminating in their learning. This means that some games may be lumped together with respect to the learning process and individuals might learn to behave some way in a group of games rather than some specific game. Consequently, some equilibria of the games are destabilized and new equilibria are created. Other models of evolution in cases of incomplete information have also been proposed where individuals with limited cognitive resources face multiple games [1,26].

3.5 Evolution of the Game

Evolution in circumstances of incomplete information is an interesting and useful tool for modeling many types of social interaction. However, uncertainty about the game is not the only important factor in the cases raised in the introduction. To understand the effects of environmental degradation on social behavior, for example, we need to model changes in the underlying game—Nature changes over time.

Suppose there is a circumstance where the game being played may shift between g_1, g_2, ..., g_n. Let G denote this set of games. For simplicity, we will begin with the case where each $g \in G$ has the same strategy set for all players. Thus, the evolution will primarily concern change in the payoffs of the game, though we will also include a brief discussion of other methods of modeling evolving games, which includes expanding or reducing the strategy space.

Following Harsanyi's method for studying games of incomplete information, we will also conceive of Nature as a "player" in the game. Nature will have as many "strategies" as there are ways the game could change. As before, we will think in terms of a population of Nature players. A population for Nature is a vector $z = (z_1 \ldots z_n)$, where $z_i \geq 0$ and $\Sigma z_i = 1$. In this representation, there are n games and each z_g corresponds to some specific game g.

Let $F_g(s, x)$ be the fitness of strategy s in population x with respect to game g. Then, we can compute the fitness value relative to a specific z (i.e., a probability distribution over the possible games) with Eq. (3.3). It is important to note that by considering Nature as a player in the game, one is free to use standard notions from game theory, such as the Nash equilibrium, in analyzing the evolving game. Equilibria in the evolving game must be equilibria in the extended game where Nature is a player. The replicator dynamics for a specific z is specified by Eq. (3.4).

$$F_G(s, x, z) = \sum_{g \in G} F_g(s, x) z_g \tag{3.3}$$

$$\frac{dx_i}{dt} = x_i \left[F_G(i, x, z) - F_G(x, x, z)\right] \tag{3.4}$$

where $F_G(x, x, z)$ is the average payoff to strategies in the player population. What is left is to introduce the ways in which z may be changing over time. To do this, we simply need a fitness function for z and a set of differential equations governing the change of z.

However, the underlying games do not specify any payoff function for our fictitious player "Nature," thus there is no unique way to determine F_G for some a given z and x. This is both an advantage and a disadvantage. The advantage is the wide degree of modeling flexibility. The disadvantage is that there is little guidance from game theory in this matter and that we must simultaneously account for the co-evolution of player behavior.

Consequently, the way in which the game will change must be specified for each model in a way that is dependent on player behavior. This can be done by explicitly introducing a set of payoffs for Nature, which will govern how the game changes as player behavior evolves. Essentially, we can create a three-player game with two interacting populations. Two players are drawn from the player population to interact along with a third player is drawn from the Nature population. Thus, Nature will receive payoffs and be able to evolve along with the players.

To do this, first we construct a strategy space for Nature S_N, which corresponds to the set of possible games G. Then, we can define a new payoff function for each game which, for the players, is the same as before, but also includes a payoff function for Nature π_N that provides a payoff given a strategy profile (i, j, g) for two players from x and a game g. Then, we define a new payoff function for each game which, for the players, is the same as before, but also specifies a payoff for Nature. Let $\pi_N(i, j, g)$ specify Nature's payoff for strategy profile (i, j) in game g.

$$F_N(x, g) = \sum_{i \in S} \sum_{j \in S} \pi_N(i, j, g) x_i x_j \tag{3.5}$$

We can then introduce an evolutionary dynamic acting on the Nature population. For simplicity, we will use the replicator dynamics for Nature as well as for the player population. For the Nature population this yields Eq. (3.6).

$$\frac{dz_i}{dt} = z_i[F_N(x, i) - F(x, z)] \tag{3.6}$$

$F_N(x, z)$ is the average payoff for the Nature population. There is, of course, no strict requirement to use the replicator dynamics to govern the evolution of z. Once F_G and F_N is specified, one could specify whichever dynamics was most appropriate for the setting they wished to model—including different dynamics for the player population and the receiver population. However, one potential advantage to using the replicator dynamic for both populations is that the connections between dynamic stability and static equilibria of the game mentioned above were established by assuming a uniform dynamic across populations [11,24]. Little is known about these connections when there are multiple types of dynamics working simultaneously.

Hashimoto and Kumagai present a model of what they call "Meta-evolutionary game dynamics" similar to the approach taken here [10]. While they do not include a fictitious Nature player, there are a number of games that players may face each with a corresponding weight, analogous to the z_i values from above. The weights are then adjusted according to a "meta-evolutionary" dynamic. The dynamic operates on various valuations assigned to the games based on the behavior of and current payoffs to the players, which may be changing dynamically as the weights on the game change. This approach could be incorporated into the model above by appropriately specifying the payoff to Nature.

There are two advantages to the explicit inclusion of the fictitious player Nature. First, this characterization makes it clear that the game may shift in ways that are not directly tied to the current or past payoffs to the players (although this is not ruled out either). Furthermore, it is not necessary that Nature's payoffs be fixed in the game matrices; for example, although we will not do this here, Nature's payoff could shift as a function of time and be used to represent games shifting in periodic oscillations. Second, by explicitly considering a population that represents the distribution of Nature, modeling evolving games becomes a straightforward extension of multi-population models and many results from those areas may be directly applicable; see [18].

3.6 Other Models of Evolving Games

The modeling framework presented above captures a dynamic change in the payoffs of the game. However, there are other ways that a game may evolve which cannot be directly expressed as a change in the payoffs. In this section, I will briefly discuss some other recent models of changing games and contrast them with the model above.

Worden and Levin present a model of an evolving game as a possible solution to the cooperation problem posed by the prisoner's dilemma [25]. In their model, new strategies can develop that are slight variants of existing strategies and have slight differences (determined by random variables) in the payoffs. Over time, these new strategies can overcome the strict dominance of defection in the prisoner's dilemma. In this model, unlike the model above, the evolving game is open-ended and the results of evolution are not deterministic. This represents a game changing gradually due to random perturbations in the strategies, while the model above represents a game changing systematically due to some determined rule or known factor.

In another case, Skyrms provides an overview of recent work on the evolution of signaling from a game-theoretic perspective [21]. One set of central questions Skyrms is seeking to answer is if and when simple reinforcement learners can develop signaling systems. He presents a model where signaling systems can emerge without any signals to begin with—learners invent them. Skyrms models invention as a random occurrence in the reinforcement learning process. With a small probability, the sender may try a brand new strategy (signal or response in this case). If those strategies are successful, they begin to reinforce on that new action. Over time, simulations reveal that the learners invent the signals they need to communicate about the world. This is an evolving game; the strategies available to the players is changing over time as the result of the learning process. The general structure of the signaling game allows for a systematic model of introducing new signals (and corresponding strategies). However, this approach may not generalize to games with other structures.

In addition to the evolution of new strategies, one may want to consider the elimination of existing strategies. For instance, Paul and Ramanujam consider a systematic elimination of strategies in large games, where society can eliminate strategies if they are unused [17]. The introduction or elimination of strategies to a game is one way that a game may evolve that cannot be directly captured by appeal to changing payoffs. Despite these limitations, the approach I have described here can be useful in modeling ways in which evolving games relate to social contracts.

3.7 Social Contracts with Evolving Games

Examples from descriptive social contract theory motivated the model of evolving games presented above. The formation of social contracts may crucially involve changing the underlying game being played. In this section we examine three simple cases relevant to the formation of social contracts and present a model in terms of evolving games for each. The first two pertain to the evolution of cooperation; the third pertains to the evolution of fairness and distributive justice. None of these models are intended as providing a comprehensive representation of these issues within the framework of changing games. Instead, these are intended as suggestive illustrations of how the model of evolving games may be applied to some of the central topics studied in the evolution of the social contract.

3.7.1 Cooperation with Environmental Degradation

Establishing cooperative outcomes in games such as the Prisoner's Dilemma and the Stag Hunt are seen as an important aspect of many social contracts [19,20]. One area of contemporary importance in this regard is achieving cooperation with respect to climate change. DeCanio and Fremstad apply game theoretic analysis to international climate negotiations [6]. The current climate situation is often represented as a type of Prisoner's Dilemma. DeCanio and Fremstad argue that given the high stakes of managing climate change, the situation might be better modeled as a coordination game (like the Stag Hunt). Let us take this idea one step further and imagine environmental degradation that is brought about by a certain type of behavior. The game may be a Prisoner's Dilemma, but become a Stag Hunt as the environment deteriorates due to uncooperative behavior.

Consider the two games from the introduction but slightly modified. To model a degrading environment and how this may impact the game, we would imagine that each possible outcome in the Prisoner's Dilemma as well as the Stag Hunt also carried a payoff for Nature, and this payoff dictates the way in which Nature changes over time. The third payoff listed in each matrix is the payoff to Nature, who is "choosing" which game to play. These games are shown in Figs. 3.3 and 3.4.

	Cooperate	Defect
Cooperate	2, 2, 1	0, 3, 0
Defect	3, 0, 0	1, 1, 0

Figure 3.3 Prisoner's Dilemma including a payoff to Nature.

	Cooperate	Defect
Cooperate	1, 1, 0	–2, 0, 0
Defect	0, –2, 0	0, 0, 1

Figure 3.4 Stag Hunt including a payoff to Nature.

In essence, this model is a three-player game where two players are drawn from a single player-population, and the fictitious player Nature is represented by a second population. Nature will evolve relative to the actions of the other population. Given the payoff structure above, Nature will shift toward the Stag Hunt if defection is prominent—the idea being that the environmental degradation makes the cooperative outcome more important to achieve. Nature will shift to a Prisoner's Dilemma if cooperation is prominent—the idea being that if little damage is done to the environment a cooperative solution is less important and the temptation to defect becomes dominant.

In this case, the shifting game causes cooperation in the Stag Hunt to become destabilized, resulting in a unique equilibrium between players and their environment: defection in the Stag Hunt. The phase portrait in Fig. 3.5 shows the replicator dynamics of the model. The *x*-axis represents the frequency Cooperators (or Defectors) in the population of players: points to the right side represent a higher frequency of Cooperators. The *y*-axis represents the bias of Nature in presenting the Stag Hunt or the Prisoner's Dilemma to the players. Points near the top represent a higher frequency of the Stag Hunt. Altering the payoffs to Nature can alter the relative speed of the evolution of the game. This is a bleak picture for the evolution of cooperation in degrading environments: even if cooperation is a possible solution in a bad environment, an evolving game may destabilize that solution.

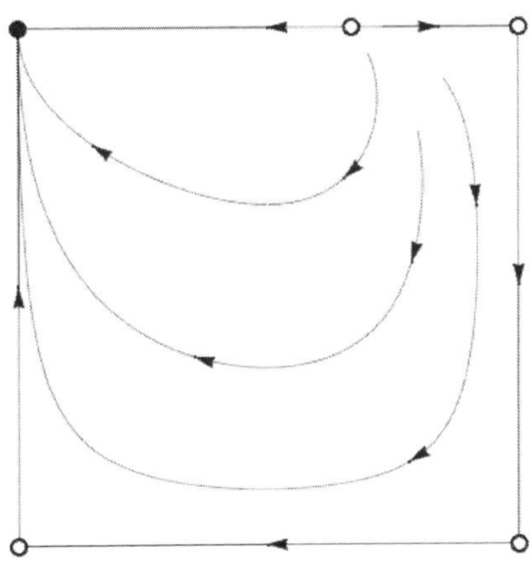

Figure 3.5 Game shifting between the Stag Hunt and the Prisoner's Dilemma. The *x*-axis represents the distribution of the player-population, further right representing higher frequency of cooperation. The *y*-axis represents the frequency with which Nature selects the Stag Hunt (higher) or Prisoner's Dilemma (lower).

3.7.2 Cultural Shifts: Increasing the Value of Cooperation

There are other ways, however, that the game may evolve and in many cases the prospects for establishing cooperation may be much better. Bicchieri provides an account of how social norms arise and how they affect the behavior of individuals in the social games they play [3]. Bicchieri's account of social norms involves various psychological elements involving beliefs about what kind of behavior others will expect, pressure to conform, etc. These psychological aspects of social norms mean that these norms can alter individual incentives. One idea that comes out of this approach is that the way individuals view the payoff structure of the game may shift due to social norms. I will not attempt to directly capture any of Bicchieri's models within the framework of this paper. Instead, I will construct a simple model that illustrates one way we can represent a cultural-shift in how the game is viewed.

When introducing the replicator dynamics above, I mentioned that there are two common interpretations: one biological, one cultural. We have, thus far, been thinking of a changing game as a change in the underlying payoff structure. However, what the payoffs represent is different in each interpretation. In the biological setting, the payoffs determine relative reproductive success. In the cultural setting, payoffs are translated into a propensity to be imitated by others.

Under the cultural evolution interpretation, a change in an underlying game could represent a change in the way the game is being perceived by the individuals playing it. If there is a cultural shift to see some outcomes as better than others, and more likely to be imitated, this is essentially a change in the payoffs of the game. Hence, we should be able to interpret a change in the underlying game culturally just as we interpret the dynamic change in the behavior culturally.

Consider Figs. 3.6 and 3.7 as an example. Imagine that a population of individuals might shift between playing a Prisoner's Dilemma and a Prisoner's Delight. Note that the only difference in the players' payoffs between the two games is the payoff to cooperation (simply + 2 in the Prisoner's Delight relative to the Prisoner's Dilemma).

	Cooperate	Defect
Cooperate	2, 2, 1	0, 3, 0
Defect	3, 0, 0	1, 1, 0

Figure 3.6 Prisoner's Dilemma including a payoff to Nature.

	Cooperate	Defect
Cooperate	4, 4, 0	2, 3, 0
Defect	3, 2, 0	1, 1, 1

Figure 3.7 The Prisoner's Delight including a payoff to Nature.

In this case, the Nature population represents how the player population views the game. Imagine that the population will tend to value cooperation more, the less there is of it in the population. Here, this shift in the value of cooperation can be represented by Nature getting a payoff for selecting the Prisoner's Delight when defection is occurring and no payoff in any other case. In this toy model, if there is a lot of cooperation, people do not change how they value cooperation. If there is little cooperation, people value it more.

The dynamics of this evolving game are represented in Fig. 3.8. The population always winds up cooperative, but the degree to which they value cooperation (i.e., the degree to which they treat the game as a Prisoner's Delight) varies depending on the starting point. Interestingly, this example is such that the lower the initial value to cooperation, the higher the resulting value to cooperation.

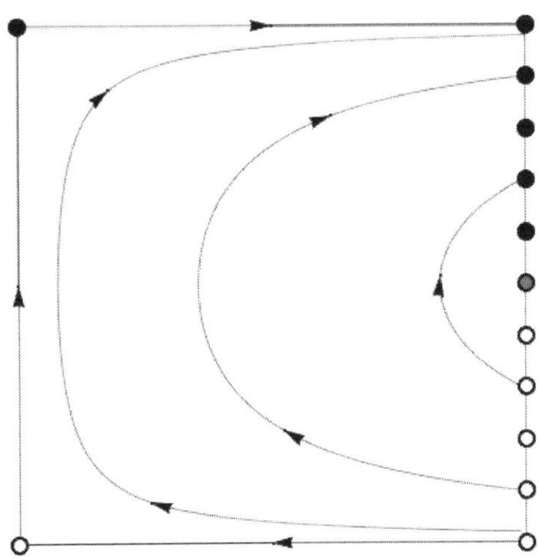

Figure 3.8 Game shifting between the Prisoner's Dilemma (bottom) and the Prisoner's Delight (top) as a result of a cultural shift in valuing cooperation (right) when defection (left) is frequent.

3.7.3 Fair Division in an Evolving Game

Thus far, we have only considered examples involving two strategy games. However, there are other important games for the evolution of the social contract. One such game is the Nash Demand game where two players must submit simultaneous demands on some shared resource [15]. Each gets their demand if they are compatible and each gets nothing otherwise. This game and other similar bargaining games have been used to model distributive justice and the evolution of fairness [5,19]. The intuitively fair solution to this game is to demand half of the available resources; this corresponds to Nash's original axiomatic solution. Figure 3.9 shows a simplified version of this game where there are six units of some resource to divide and each can demand 2/3, 1/2, or 1/3.

	Greedy	Fair	Modest
Greedy	0, 0	0, 0	4, 2
Fair	0, 0	3, 3	3, 2
Modest	2, 4	2, 3	2, 2

Figure 3.9 Simplified Nash Demand game.

Evolution in this simplified Nash Demand game will lead the population to one of two evolutionarily stable states: Everyone plays "Fair" or half of the population plays "Greedy" and the other half plays "Modest" [19]. The second state is inefficient: encounters between two Greedy individuals result in conflict and encounters between Modest individuals result in less than the whole resource claimed. Skyrms refers to this mixed population as a "polymorphic trap" and views this state as an obstacle to the evolution of fair behavior [19]. This led Skyrms to consider mechanisms of correlated interaction that may avoid the polymorphic trap and enhance the evolution of fair behavior. Along similar lines, the evolving games framework above can be used to examine other mechanisms that may make fair behavior more likely to evolve.

Let's consider one possible way this game could evolve. Suppose that persistent inefficient outcomes lead to a new game with the same structure but with reduced payoffs for successful demands. Specifically, positive (non-zero) payoffs are reduced by 2. This could happen for a number of reasons. For example, a society could adopt a policy of taxing gains from certain exchanges when there are inefficiencies present. Alternatively, this may happen without a governmental policy prompting the shift if worsening conditions due to inefficiency mean that small payoffs are no longer considered acceptable and are not strictly preferred to zero payoff.

Nature will shift toward the standard game when efficient outcomes occur and will shift toward the modified game (with –2 to positive earnings) when inefficient outcomes occur. Notice that Nature only responds to efficiency and not to fairness. More specifically, Nature will receive a payoff of 1 toward the standard game when efficient outcomes occur and a payoff of 2 toward the modified game when inefficient outcomes occur.

We now have a case where this three-strategy bargaining game will evolve as play unfolds. This game is more difficult to visualize than the two-strategy games previously considered, but can be easily explored with numerical computer simulations. Simulations, written using C, employed the discrete time version of the replicator dynamic [24]. There were 10,000 simulations, each initialized with a random distribution of strategies for both the player population and Nature. Every simulation resulted in Nature playing the standard game 100% of the time and the player population uniformly adopting the "Fair" strategy. All simulations reached this approximate state within 2,000 generations.

These results show that if everyone is punished for inefficiencies in the Nash Demand game, then fair division can result. This example illustrates another possible application of the evolving games framework. If a social planner, say, wanted to avoid certain suboptimal social equilibria, one could consider possible policy changes in terms of evolving games. Of course, this is but one of many ways that this game could evolve. And there may be many other important aspects to consider. This example serves only to illustrate how we can apply the framework above to represent possibilities that are difficult to capture without a model of evolving games.

3.8 Conclusion

It is possible to represent some important aspects of an evolving game within existing tools of evolutionary game theory. The above framework allows us to model a dynamic change in the payoffs of the game, which may be dependent on the actions of the players. This is important for modeling the evolution of the social contract as many aspects of social contracts involve the adoption of rules, norms, or institutions that change the game being played. We considered three examples of the way in which games of cooperation or fair division may evolve that are relevant to the social contract. These examples are intended to be illustrative and are by no means exhaustive. Nonetheless, certain aspects of changing games that are relevant to the social contract can be easily represented as evolving games.

However, it is important to note that the modeling approach explored here also has some significant limitations and there are ways in which the model could be fruitfully extended. For example, we have used the replicator dynamics for both the player and Nature populations, but it is unlikely that Nature will change in the same way that populations do. Consequently, using the same dynamics for both Nature and players should be considered only a special case and other dynamics for one or both populations will be important to consider. Non-deterministic dynamics may also be examined. This can be done either by simply introducing some kind of noise into the evolutionary dynamics or by examining other kinds of evolutionary processes, e.g., genetic algorithms.

Another limitation of the modeling framework used here is that it does not explicitly include the possibility of evolving new strategies, or strategies being eliminated, over time. The discovery of novel strategies is particularly difficult to model using traditional game theoretic tools because new strategies can change both the payoffs and the structure of a game. Expanding the model to include these possibilities is important for a complete representation of the ways in which games evolve.

Finally, this model of evolving games involves substantial idealizing assumptions such as randomly mixing and infinite populations. For these reasons, it would be of interest to employ agent-based modeling tools to extend the model of evolving games. Agent-based models can explore complex interaction settings with heterogeneous individuals where learning is the result of stochastic processes. Such models would be more complex, but would also allow for a more realistic representation of settings where games are evolving. Despite the differences between agent-based models and the population-models explored here, there are some aspects that can be readily extended to agent-based models. In particular, the representation of an evolving game by changes of a fictitious player Nature that is "choosing" the game does not require the population-dynamics approach used here. This same general method of representation could also be extended to agent-based models. Such extensions are a promising area for future research.

Acknowledgments

I would like to thank Brian Skyrms, Patrick Forber, and the members of the 2012 Complexity and the Human Experience conference for helpful comments and suggestions.

References

1. Bednar, J., and Page, S. (2007). Can game(s) theory explain culture? The emergence of cultural behavior within multiple games. *Rationality Soc*, **19**, 65–97.
2. Bergstrom, C. T., and Godfrey-Smith, P. (1998). On the evolution of behavioral heterogeneity in individuals and populations. *Biol Philos*, **13**, 205–231.
3. Bicchieri, C. (2006). *The Grammar of Society: The Nature and Dynamics of Social Norms.* (Cambridge University Press, New York).
4. Binmore, K. (2004) The Breakdown of Social Contracts. In *Social Dynamics* (ed. Durlauf, S. N., and Young, H. P.), Chapter 8, MIT Press, Cambridge MA, 213–234.
5. Binmore, K. (2005). *Natural Justice.* (Oxford University Press, USA).
6. Decanio, S. J., and Fremstad, A. (2011). Game theory and climate diplomacy, *Ecological Economics,* 10.1016/j.ecolecon.2011.04.016.
7. Ely, J. C., and Sandholm, W. H. (2005). Evolution in Bayesian games I: theory, *Games Econ Behav*, **53**, 83–109.
8. Fudenberg, D., and Tirole, J. (1991). *Game Theory.* (The MIT Press, Cambridge MA).
9. Harsanyi, J. (1967). Games with incomplete information played by Bayesian players, *Manag Sci*, **14**, 159–182.
10. Hashimoto, T., and Kumagai, Y. (2003). *Advances in Artificial Life* (ed. W. Banzhaf, W., Christaller, T., and Ziegler, J.), Meta-evolutionary

game dynamics for mathematical modeling of rules dynamics (Springer) pp. 107–117.

11. Hofbauer, J., and Sigmund, K. (1998). *Evolutionary Games and Population Dynamics* (Cambridge University Press, USA).
12. Huttegger, S. M., and Zollman, K. J. S. (2010). Dynamic stability and basins of attraction in the sir Philip Sidney game, *Proc. R. Soc. B,* **277**, 1915–1922.
13. Kollock, P. (1998). Social dilemmas: the anatomy of cooperation, *Ann Rev Sociol,* **24**, 183–214.
14. Mengel, F. (2012). Learning across games, *Games Econ Behav,* **74**, 601–619.
15. Nash, J. (1950). The bargaining problem, *Econometrica,* **18**, 155–162.
16. Ostrom, E. (1990). *Governing The Commons: The Evolution of Institutions for Collective Action* (Cambridge University Press, USA).
17. Paul, S., and Ramanujam, R. (2011). Dynamic restriction of choices: synthesis of societal rules. In *Logic, Rationality, and Interaction* (ed. Van Ditmarsch, H., Lang, J., and Shier, J.), Springer, pp. 28–50.
18. Sandholm, W. H. (2010). *Population Games and Evolutionary Dynamics* (MIT Press, Cambridge MA).
19. Skyrms, B. (1996). *Evolution of the Social Contract* (Cambridge University Press, USA).
20. Skyrms, B. (2004). *The Stag Hunt and the Evolution of Social Structure* (Cambridge University Press, USA).
21. Skyrms, B. (2010). *Signals: Evolution, Learning, and the Flow of Information* (Oxford University Press, USA).
22. Taylor, P., and Jonker, L. (1978). Evolutionary stable strategies and game dynamics, *Math Biosci,* **40**, 145–56.
23. Thomas, B. (1984). Evolutionary stability: states and strategies, *Theor Popul Biol,* **26**, 49–67.
24. Weibull, J. W. (1995). *Evolutionary Game Theory* (MIT Press, Cambridge MA).
25. Worden, L., and Levin, S. A. (2007). Evolutionary escape from the prisoner's dilemma, *J Theor Biol,* **245**, 411–411.
26. Zollman, K. J. S. (2008). Explaining fairness in complex environments. *Polit, Philos Econ,* **7**, 81–98.

Chapter 4

Character Networks for Narrative Generation: Structural Balance Theory and the Emergence of Proto-Narratives

Graham Alexander Sack

Department of English and Comparative Literature, Columbia University, Philosophy Hall, New York, NY 10027, USA

gas2117@columbia.edu

This chapter models narrative as a complex adaptive system in which the temporal sequence of events constituting a story emerges out of cascading local interactions between nodes in a social network. The approach is not intended as a general theory of narrative, but rather as a particular generative mechanism relevant to several academic communities: (1) literary critics and narrative theorists interested in new models for narrative analysis, (2) artificial intelligence researchers and video game designers interested in new mechanisms for narrative generation, and (3) complex systems theorists interested in novel applications of agent-based modeling and network theory. The chapter is divided into two parts. The first part offers examples of research by literary critics on the relationship between social networks of fictional characters and the structure of long-form narratives, particularly novels. The second part provides an example of schematic story generation

Complexity and the Human Experience: Modeling Complexity in the Humanities and Social Sciences
Edited by Paul A. Youngman and Mirsad Hadzikadic

ISBN 978-981-4463-26-3 (Hardcover), 978-981-4463-27-0 (eBook)
www.panstanford.com

based on a simulation of the structural balance network model. I will argue that if literary critics can better understand sophisticated narratives by extracting networks from them, then narrative intelligence researchers can benefit by inverting the process, that is, by generating narratives from networks.

4.1 Introduction

Throughout this chapter, I will make extensive use of concepts from social network analysis and structural balance theory. The basic unit of analysis in structural balance theory is the *triad*, defined as a triangular configuration of friendship and enmity ties between three mutually connected nodes. Some triad configurations are socially unstable and, when embedded in networks with many interdependencies, may trigger cascading social events. These cascades, I will argue below, can be treated as a proto-narrative—the skeleton of a story from which complex social dramas may be constructed.

My approach to narrative is loosely inspired by a variety of sources.

In *Deceit, Desire, and the Novel* (1961), literary critic Rene Girard argues that a defining feature of the novel as a modern story-telling form is the way characters' desires are embedded in and mediated by indirect social relations. Taking Cervantes' *Don Quixote* as the prototype for the modern novel, Girard asserts that Quixote's desire is "triangular":

> The straight line is present in the desire of Don Quixote, but it is not essential. The mediator is there, above that line, radiating toward both the subject and the object. The spatial metaphor that expresses this triple relationship is obviously the triangle [10].

Girard argues that in novels such as *Don Quixote*, the subject desires the object not because of its inherent qualities, but because some third character, the mediator, also desires the object. The mediator may be a role model who the subject intentionally imitates, or he may be a character against whom the subject competes for the desired object, as with a rival in polite society, romance, or commerce. In either case, the dyadic relationship between subject and object cannot be understood without reference to the mediator. While I will not use the concept of "triangular desire" directly, I will adapt Girard's general interpretive framework by drawing a parallel between the idea that *triads of characters* are the basic unit of *narrative* analysis and structural balance theory's assumption that *triads of nodes* are the basic unit of *network* analysis.

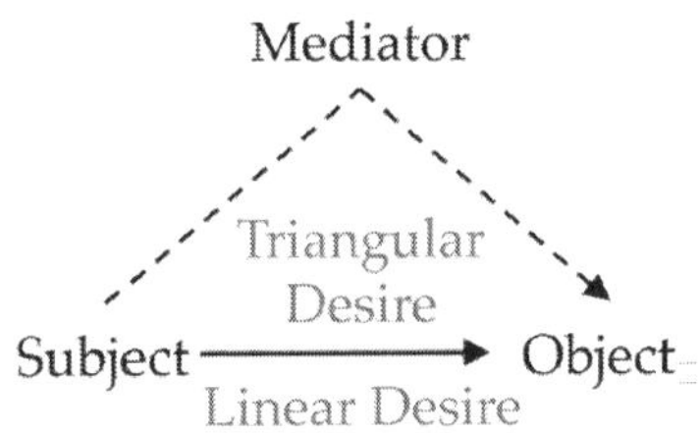

In *Identity and Control: How Social Formations Emerge* (2008), network theorist Harrison White posits, "social networks emerge only as ties mesh with stories" [25]. White suggests that social ties ought to be thought of multi-dimensionally in terms of "netdoms": "'dom' from domain of topics and 'net' from network relations." When two social agents encounter one another, they struggle for recognition and status by "switch[ing] from netdom to netdom, finding footings in different networks in differing domain contexts." Two co-workers, for example, might initially relate to one another professionally, but then switch between political, religious, or even romantic domains as the social tie between them evolves. The more "netdom switchings" occur, the more complex and nuanced the relationship between the identities becomes. Over time, these switchings settle down into a stable tie that is comprehended by the participants via a "story." White summarizes the process as follows:

> A story is a tie placed in context. Stories structure switchings into accounts with a beginning, middle, and end; so story-making frames social time.... These relations are characterized by stories told in and about them with meanings drawn from the switchings between netdoms.... A network can be traced as similar stories appear across a spread of dyads [25].

White's notion of "netdom switchings" provides another potential link between structural balance theory and narrative. As the unstable triads in a structural balance network evolve toward stability, the edges connecting each pair of nodes undergo a simplified version of domain switching, oscillating between friendship and enmity. Following White's logic, the more frequently a link in a triad switches, the more complex and nuanced the relationship between the associated node-characters becomes, gradually forming a story-tie.

I instantiate the structural balance model in NetLogo, an IDE for agent based modeling, and run the simulation forward in time under different parameter configurations, producing a range of possible proto-narrative event sequences that vary in length and outcome. My use of simulation to generate proto-narratives is influenced by the work of computational social scientists Joshua Epstein and Robert Axtell. In *Growing Artificial Societies* (1996), the authors describe how their now canonical "Sugarscape" model, originally constructed to study the emergence of wealth inequality, can generate "proto-histories"—schematic social and cultural histories in which individuals agglomerate and form tribes, battle, and trade. Although they do not use the term, Epstein and Axtell argue, in effect, that each run of the Sugarscape model is analogous to an *historical narrative.*[a] Although simulations such as Sugarscape are now widely used in the social sciences, practitioners rarely if ever discuss the idea that the temporal progression of a simulation can be treated as a narrative. Epstein and Axtell emphasize moreover that they "grow this history 'from the bottom up'" [2]. That is, their proto-historical narratives are complex adaptive systems displaying the property of emergence.[b] Structural balance models provide a similarly CAS-based approach to narrative generation. Unstable triads update based on local stability rules, yet produce a narrative chain of events tracing the formation of global network structures (see the discussion of "social mitosis" below).

[a]In a chapter subtitled "The Emergence of History," Epstein and Axtell write, "The basic aim of this chapter is to "grow" a very simple caricature of history—a "proto-history" if you will.... The social story is as follows: In the beginning, there is a small population of agents..." [2].

[b]"Emergent narrative" is an overloaded term. Researchers in interactive narrative use the phrase, along with "character based narrative," to refer to minimally plotted stories generated spontaneously through live user interaction (See Aylett 1999 [3]). My meaning is drawn from complex systems theory and refers to the emergence of systemic properties from local interactions. A connection could be drawn—one might argue that human interactions with non-player characters in virtual reality environments exhibit CAS-like behavior—but the terms are not trivially synonymous.

4.2 From Narratives to Networks

Over the past several years, literary critics have begun researching the relationship between social networks and narrative structure, including several efforts to extract character networks from literary works [7,19,21]. The guiding principle behind literary network analysis is that narratives are not merely depictions of individual experience in language but are also *artificial societies* whose imaginary social forms can be quantified and analyzed. What such analyses reveal is that *narrative* structure, such as plot, genre, and characterization, is intimately related to *network* structure.

Included below are examples drawn from my own research on literary networks. Figure 4.1 shows the sociograms[c] for several canonical European novels: Cervantes's *Don Quixote de la Mancha* (1605), Charles Dickens's *David Copperfield* (1850), and Virginia Woolf's *Mrs. Dalloway* (1925). Noticeable contrasts between these networks reflect key differences in literary conventions across historical periods and genres.

Don Quixote, widely regarded as the first modern novel, possesses an episodic plot structure derived in part from the *picaresque* story-forms popular during the Spanish Golden Age. It consists of a series of adventures that all feature the iconic knight and his squire but which are otherwise minimally connected in terms of plot. *Don Quixote*, moreover, operates as a frame narrative, encompassing many interpolated stories-within-the-story—such as Cardenio's autobiography or the pastoral poems describing Marcela. The sociogram reflects this disconnected plot structure. The network is centered on a main axis connecting Don Quixote and Sancho, which is embedded in a diffuse web of characters that extends outwards in several layers. For a relatively small network, it has a high diameter, indicative of the fact that it is effectively a *network of networks*: When Don Quixote meets another character, such as Cardenio, often this new acquaintance temporarily assumes the role of narrator, relating a micro-narrative with its own stand-alone character network not linked to the original action. We can see this most clearly in the presence of several peripheral cliques,

[c]Social networks are generated for each novel by running a 10-word window through the text and counting the number of times each pair of character names co-occurs. Edges are drawn for pairs with ≥3 co-occurrences and are weighted by frequency.

particularly the one between Anselmo, Lothario, Camilla, and Leonela, whose tale appears in a found manuscript read aloud by Quixote's priest. Moreover, the network contains a low proportion of "strong ties": Aside from the heavily weighted edge between the knight and his squire, most edges are thin and light, indicating brief, glancing interactions with secondary characters who provide color and variety in particular episodes but who rarely recur or interact significantly with one another.

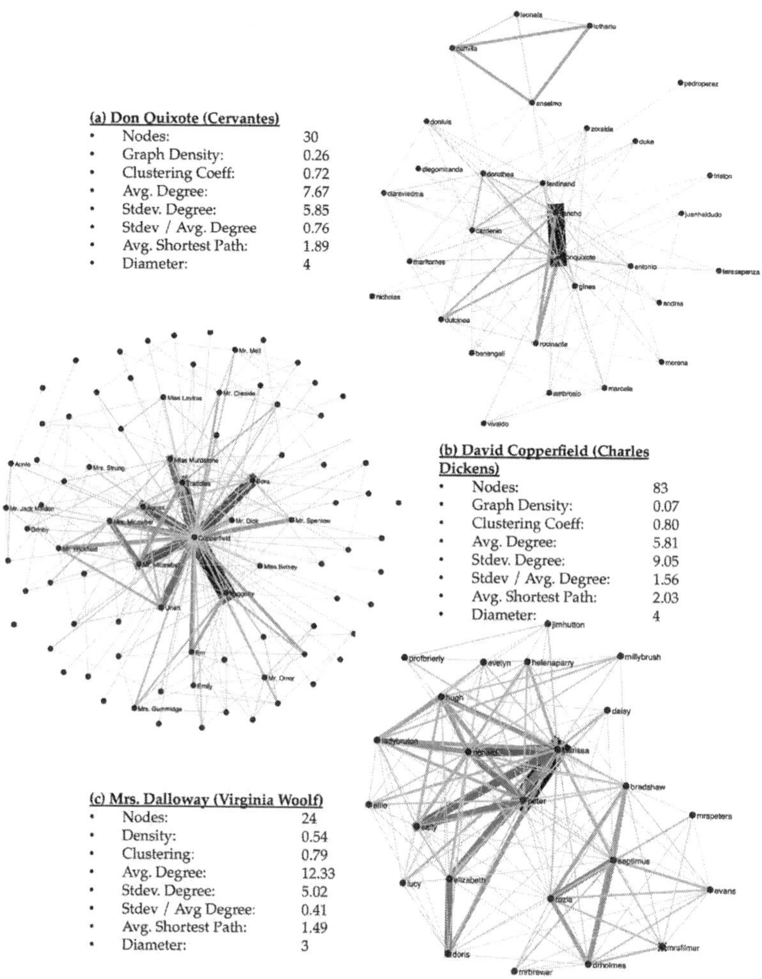

Figure 4.1 Character network sociograms for three illustrative novels.

David Copperfield is one of the preeminent examples of the 19th Century *Bildungsroman*, depicting the education and development of its eponymous protagonist as he finds his way in the world, seeking benefactors, a career, and a marriage partner. Like many serialized mid-Victorian novels, it features an expansive cast: The network consists of 83 nodes, with a high proportion of isolates and low graph density (7%). The network is highly centralized with an obvious star-shape, reflecting an egocentric focus on its protagonist, who is the hub for virtually all character interactions. This accounts for its very high standard deviation in node degree, indicative of inequality in connectedness and social importance. The prevalence of strong ties is noticeably greater than in *Don Quixote*, illustrative of the 19th Century *Bildungsroman*'s concern with complex relationships developed over a long duration, rather than the brief encounters with strangers common in the Renaissance-era *picaresque*.

The character network for *Mrs. Dalloway* contrasts noticeably with the others. While mid-Victorian novels often featured sprawling casts, the network for *Mrs. Dalloway*, a canonical work of high British modernism, is delimited. The focus is on psychological depth rather than sociological breadth. The sociogram consists of a single large component with no isolates and a very high graph density (54%) and clustering coefficient (79%), reflecting embedded relationships between characters with many common social ties. Unlike *David Copperfield*, the narrative is not singularly focused on its title character: Point of view shifts approximately every ten pages. The network, correspondingly, does not have a pronounced center: It is bifurcated into two cliques–one concentrated around Clarissa Dalloway and the other around her narrative alter ego, Septimus. Moreover, the network has a low standard deviation in node degree, indicating that character interaction is broadly and evenly distributed across the ensemble. Lastly, the diagram exhibits a high proportion of strong ties. The overall picture is that of a tightly knit social world focused on the intimate relationships between a small set of equally significant characters.

As these brief examples suggest, there is a close association between narrative structure and network structure. Authorial decisions related to linear versus episodic plot or the balance of focus between protagonist and ensemble are visible in network properties such as centralization, graph density, diameter, clustering, and prevalence of strong versus weak ties. But if literary critics can better understand sophisticated narratives by extracting networks from them, perhaps narrative intelligence researchers can benefit by inverting the process, that is, by generating narratives from networks.

4.3 From Networks to Narratives

In the remainder of this chapter, I offer a simple example of how social networks may be used to generate narratives. While the networks in the preceding section *summarize* character interactions, the networks that follow *produce* interactions. Descriptive and generative networks may at first appear quite different, but I will relate them by showing how structural balance networks produce event sequences that can then be converted back into descriptive sociograms analogous to those shown above. The direct connection will be established toward the end of this section.

4.3.1 Background: Structural Balance Model

The model I will describe is based on ideas from structural balance theory, also known as social balance theory. SBT was originated in the mid-1940s by Fritz Heider, who studied patterns of belief coherence in individual psychology [11]. In the mid-1950s, Cartwright and Harary generalized Heider's theory of coherence and applied it to social relations, representing stable and unstable configurations with basic graph theory [4]. SBT has since become a sub-branch of social network theory.

Consider a set of nodes representing, for example, people or countries. Each node may be joined to each other node by an edge, which represents their relationship. If two nodes are joined, they are either (1) friends or (2) enemies. The fundamental unit of analysis in SBT is a triad of three mutually linked nodes. A triad is considered *unstable* if there is social pressure to change one of the relationship links. It is considered *stable* if there is no social pressure to change.

Let (+) represent friendship and (–) represent enmity. As shown in Fig. 4.2a, there are several possible configurations:

(1) (+)(+)(+): If all three nodes are friends/allies, the triad is considered stable.

(2) (−)(−)(−): If all three nodes are enemies, the triad is unstable, since two nodes have an incentive to ally against the third (thereby becoming friends with each other).

(3) (+)(+)(−) or (+)(−)(+) or (−)(+)(+): If one node is friends with two that are enemies with one another, it will be pressured to pick a side, and therefore the triad is unstable.

(4) (+)(−)(−) or (−)(+)(−) or (−)(−)(+): If two nodes are friends with each other and both are enemies against a third, the triad is stable.

Cartwright and Harary summarize the rules for stability as follows: A triad is stable if the multiplicative product of the signs is positive [4]. The stability of the various triads conforms to the following simplified social principles: (1) My friend's friend is my friend; (2) my friend's enemy is my enemy; (3) my enemy's friend is my enemy; (4) my enemy's enemy is my friend.

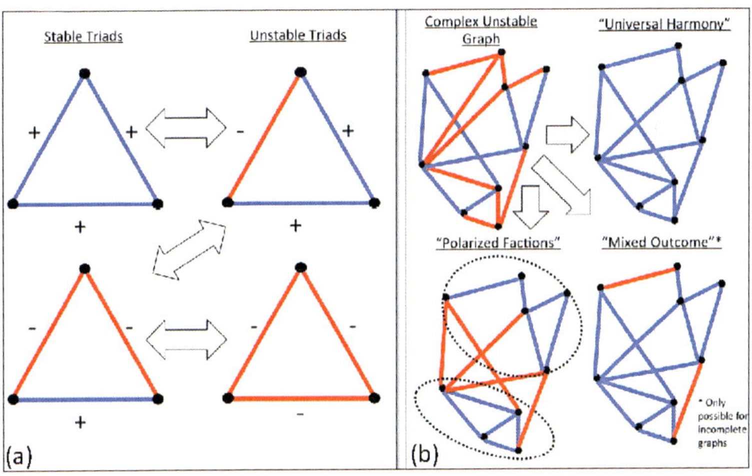

Figure 4.2 (a) Local triad stability rules. (b) Emergent globally stable network patterns.

As shown in Fig. 4.2a, changing any single link in an unstable triad will make it stable: (−)(−)(−) becomes any cyclical permutation of (−)(−)(+); (+)(+)(−) becomes either (+)(+)(+) or any cyclical permutation of (−)(−)(+). Likewise, changing any link in a stable triad will make it unstable: (+)(+)(+) becomes any cyclical permutation of (−)(+)(+); (−)(−)(+) becomes either (−)(−)(−) or any cyclical permutation (−)(+)(+).

The stability of three mutually connected nodes is easy enough to evaluate, but the complexity increases as nodes are added to create larger graphs with many interdependent triads. Nevertheless, global patterns emerge from the local interactions. One such pattern is *social mitosis.* It can be proven that there are only two ways for a complete graph, i.e., one with no missing edges, to be structurally balanced: (1) Everyone is friends (*universal harmony*). (2) There are two factions of friends with total enmity between them (*bi-polar factions*).[d] For an incomplete graph, two more outcomes are possible: (3) Nodes divide into three or more groups with total enmity between them (*multi-polar factions*). (4) some nodes are enemies but no factions form (*mixed outcome*).[e] While the general properties of the equilibrium state of any graph are deterministic, the dynamic process by which that graph reaches an equilibrium is not. This is what makes it interesting and useful as a narrative generation mechanism.

4.3.2 Model Implementation

The next several pages describe a version of the structural balance model that I have implemented in NetLogo,[f] an IDE for agent-based modeling. This is not the first computer simulation of structural balance dynamics (see [12,13,24] for alternative implementations). The crucial difference is that my focus is not on structural balance in its own right, but rather on motivating a series of observations about the narrative generating potential of social-network-based simulations. Toward this end, my emphasis is on the proto-

[d]For a simple proof of this theorem, see Chapter 5 of D. Easley and J. Kleinberg's *Networks, Crowds, and Markets: Reasoning about a Highly Connected World* (Cambridge UP, 2010).

[e]A complete graph with n nodes, has ${}_nC_2 = n(n-1)/2$ edges, each of which can be in 2 states, (+) or (−); thus, there are $2^{n(n-1)/2}$ states for the network. It can be shown that $2^{(n-1)}$ of these are stable outcomes for n odd and $2^{(n-1)} + (n!)/[(n/2)!^2]$ for n even. This corresponds to the number of ways to divide a group with n members into two factions of size m and $(n - m)$. The non-polarized solution ("universal harmony") is simply the trivial solution where $m = 0$. An incomplete graph has more stable configurations since multi-polar and mixed outcomes are permitted.

[f]To obtain a copy of the model, please email gas2117@columbia.edu.

narratives generated by the model's dynamics, which are captured by the event history, node history, and relationship-link history discussed below.

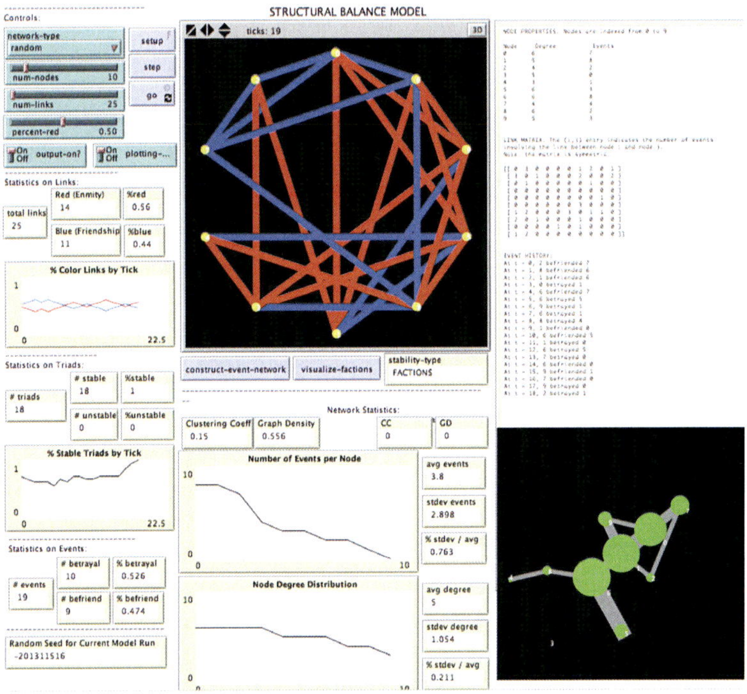

Figure 4.3 Model Interface: The control panel (upper left corner) allows the user to specify topology, number of nodes and links, and percentage of friendship vs. enmity ties in the initial structural balance network (upper center). As the model runs, link colors change generating an event history (right side), link and triad statistics (middle left side), and network statistics (lower center). Once the network reaches stability and the model halts, a character interaction network is generated from events between nodes (lower right corner).

At set-up, the user specifies the number of nodes, the number of links, the percentage of links that will be enmity, indicated in red, as opposed to friendship, indicated in blue, and the network's degree distribution, which may be uniform, random, or maximally unequal (see Fig. 4.3). With each time step, the model's algorithm checks whether there are any unstable triads. If so, the algorithm randomly selects one of the unstable triads and randomly changes one of its links from red to blue or from blue to red. Changing a link's color stabilizes the selected triad, but may inadvertently destabilize other triads. The model continues stepping forward in time until all triads in the graph have been made stable.[g]

As it runs, the model generates several types of output.

First, global network statistics, such as *graph density* and *clustering coefficient*. Since clustering coefficient measures the prevalence of complete triads, the greater its value, the more complex the structural balancing problem and the more time-steps on average required to reach stability. Second, the model tracks link and triad statistics. These include (1) number and percentage of friendship versus enmity ties and (2) number and percentage of stable versus unstable triads. The program halts when the percentage of stable triads equals 100%.

These metrics merely provide basic information about the state of the network. Of greater relevance for narrative generation are the outputs involving events, nodes, and relationships.

An "event" is defined as a change in link color. There are two types of events: (1) *Befriending*: when a red link changes to blue, meaning that the two end-nodes have changed from enemies to friends; (2) *Betrayal*: when a blue link changes to red, meaning that the two end-nodes have changed from friends to enemies. One event occurs each time step until the network reaches global stability. Each event is logged in the *event history* and is listed as "At $t = T$, node X befriended/betrayed node Y." As the model runs, it produces a simple *proto-narrative*, represented by the list of events that has occurred up to the current time step. This proto-narrative is akin to the "proto-histories" that Robert Axtell generates with the Sugarscape model.

The network topology constitutes a *rudimentary setting* representing social space rather than physical space. Like the

[g]As noted above, an n-node complete graph has $2^{(n-1)}$ stable outcomes. Because the simple algorithm implemented here proceeds through the random selection of unstable triads and the random flipping of links, there is a non-zero probability of transition between any two graph states. Therefore, the algorithm will halt in finite time. In alternative implementations, this is not necessarily so. For example, if we require that links only be flipped if doing so will immediately increase the number of stable triads, then it is possible for the graph to reach a "jammed state." For a discussion of such results, see T. Antal, P. L. Krapivsky, and S. Redner, Dynamics of social balance of networks, *Phys. Rev. E.*, **72**(3) (2005).

setting of a novel, the geometric configuration of nodes and links defines the environment in which character interaction-events will unfold. Different settings engender different event sequences. The degree distribution, representing how equal versus unequal the initial allocation of social ties is, is a key topological feature. The model has three settings: (1) uniform degree distribution, (2) random, and (3) maximum degree inequality. Maximum inequality networks are more centralized and have higher clustering coefficients given the same number of links: Because the structural balancing problem is more complex, the event history is longer on average for these networks, as shown in Fig. 4.4.

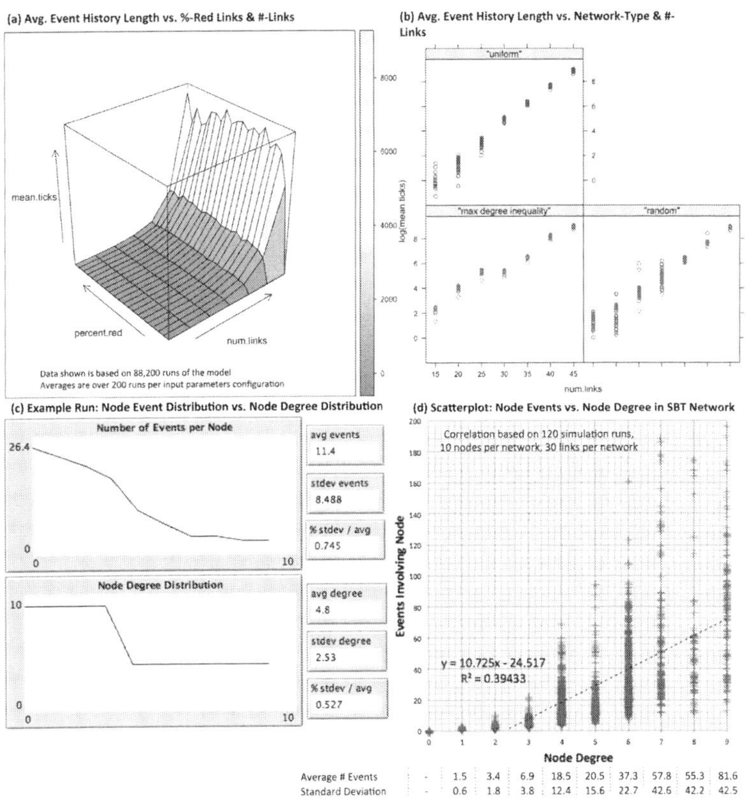

Figure 4.4 The length of the "event history," measured by the mean ticks to reach global stability, is affected by the SBT network's topology and number of links and the percentage of friendship vs. enmity ties.

Nodes may be thought of as rudimentary characters. Like characters, they have basic descriptive attributes including their degree, initial number of friends and enemies, and location in the network's topology. Nodes undergo a simple version of "character development." As the model runs, each node accumulates a personal history consisting of the link-switching events in which it has been involved. Levels of development vary. Some character-nodes are important to the proto-narrative and are involved in many events, while others are marginal.

Knowing how the number of narrative events involving a node is affected by the topological properties we assign to it enables us to place characters in the SBT network strategically, generating a desired level of character development in the final narrative. Figure 4.5 depicts the statistical relationship. As the model runs, it creates two distributions: (1) the degree distribution of the nodes at set-up, and (2) the time-evolving event distribution, showing how skewed the history of the model has been towards particular node-characters. These distributions are imperfectly correlated. The right side of Fig. 4.5 shows a scatterplot of node events vs. node degree. As we increase the degree of a node in the SBT network, the number of narrative events involving it increases. For each additional link we add, the node appears in an average of 10.7 additional narrative events (see regression equation). However, the relationship between degree and number of events is not deterministic: Node degree only explains about 40% of the variation in "character development." Other explanatory factors include local properties (how many complete triangles is the node a part of) and global properties (how long did it take for the network as a whole to reach stability). Thus, high degree nodes, called "hubs," generally figure in more events; however, it is theoretically possible to have hubs that are embedded in only stable triads. Such nodes would be central to the network, but peripheral to its narrative of development.

Link-switchings constitute rudimentary character interactions and are designated as either "befriending" or "betrayal." Just as each node in the structural balance network has a history, so does each relationship-link. Some relationships are active and tumultuous, with many oscillations between friendship and enmity, while others are uneventful. The history of each relationship is stored in an adjacency matrix where the *(i,j)* entry tracks the number of events between node-characters *i* and *j*. It can be visualized as a separate character-interaction network. Fig. 4.5 shows an example.

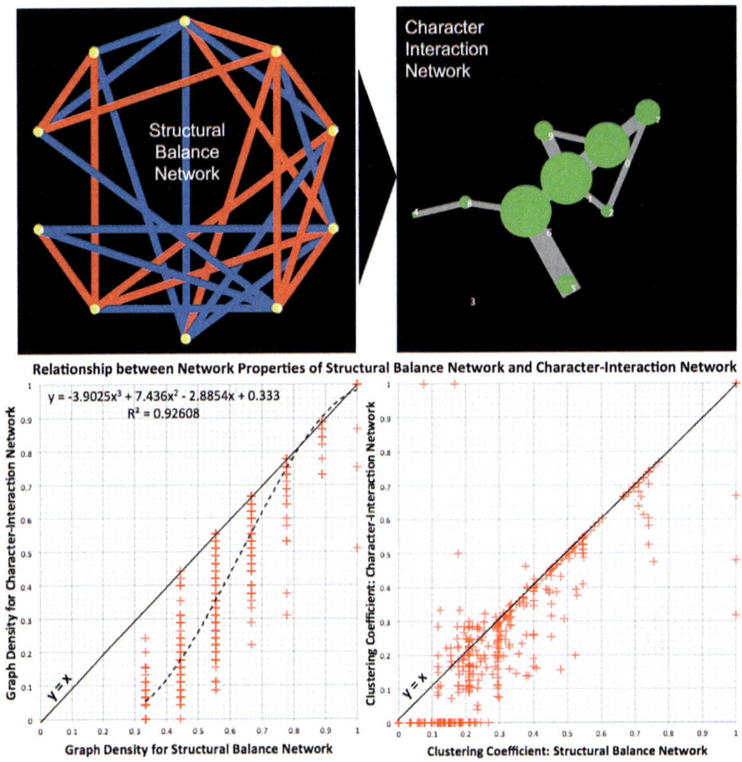

Figure 4.5 The structural balance network generates character interaction events via link-switchings. The number of events between each pair of characters is represented in a separate character-interaction network.

It is important to distinguish between the social balance theoretic network (SBT), which *generates* character interactions, and the Character Interaction network (CI), which *summarizes* them after the fact. Consistent with White's concept of "netdom switchings," as links in the structural balance network switch between friendship and enmity domains, they build up story-ties in the character interaction network. The more domain switchings in the SBT model, the stronger the tie formed. The character interaction network is comparable to the novelistic networks shown in the first half of this paper, which also summarize character interactions.

The generative SBT and descriptive CI networks may have significant topological differences: in the example model run illustrated in Fig. 4.5, the character interaction network is more centralized and has a lower clustering coefficient and graph density than the structural balance network that generated it. Understanding the relationship between the topologies is important to making effective use of the model for narrative generation. The two graphs in Fig. 4.5 provide some insight. The graph on the left shows a scatterplot of the graph densities of the SBT vs. CI networks and the graph on the right shows a scatterplot of the clustering coefficients. We can generalize a few stylized facts.

First, as the graph density and clustering coefficient of the SBT network increase, so too do those of the Character Interaction network. This is consistent with intuition: if we want our final narrative to include more connections between characters, then we should include more connections in the SBT network at set-up.

Second, the graph density of the Character Interaction network is capped by that of the SBT network. This also makes sense. The CI network can only have as many links as the SBT network that generated it. If two characters are not connected in the generative network, there cannot be any narrative events between them and therefore they will not be linked in the descriptive CI network. On the other hand, two characters being linked in the generative graph does not guarantee that they will be connected in the Character Interaction network: it merely allows for the possibility. Unlike graph density, the clustering coefficient of the CI network is not capped by that of the SBT network. It is possible for the CI network to be either more or less cliquish than the SBT network that generated it.

Third, the CI network is more likely to mirror the topology of the SBT graph the more connected the SBT graph is. We can see this from the fact that the covariance between the network features increases as the graph density and clustering coefficient of the SBT network increase: notice that the curve of best fit for graph density approaches the line $y = x$. The reason is that the more connected the SBT graph is, the thornier the problem of balancing it, and the longer the simulation will have to run to find an equilibrium solution. But the longer the simulation has to run, the more likely it is that every link will be flipped at least once, producing a fully connected Character Interaction network. This is a very helpful lever to have at our disposal when applying the model. If we want the narrative produced to have more variability and play, we can use a less connected SBT graph to generate the events. If we want to be more certain of the properties of the narrative, we can make the SBT graph more connected.

Lastly, the model's progression from instability to stability provides both a *rudimentary narrative arc* and *sense of closure*. To paraphrase Aristotle's *Poetics*, a narrative begins when an initially stable situation is disturbed by an inciting incident. The resulting disequilibrium constitutes the central problem of the narrative, toward which all actions aim according to Aristotle's principle of unity. The narrative ends when the problem is resolved, providing a sense of closure through the establishment of a new equilibrium.

The SBT simulation conforms to the Aristotelian structure, telling the simple but meaningful story of how a community journeys from an initially unstable configuration to a stable configuration. It possesses a clear beginning, middle, and end and achieves narrative closure through the establishment of a structurally balanced equilibrium. As illustrated by Fig. 4.6, each event either advances the proto-narrative toward resolution by increasing the percentage of stable triads or constitutes a reversal by decreasing the percentage of stable triads.

The SBT dynamics adhere most closely to an Aristotelian paradigm when we begin the simulation from a state only slightly perturbed from equilibrium by, for example, introducing a single friendship link between two enemy factions or a single enmity link into a network of universal friendship. These initial conditions have obvious analogues in classical narrative. The former roughly corresponds to the inciting incident in a drama such as *Romeo & Juliet*, in which members of rival factions fall in love, while the latter roughly corresponds to the inciting incident in a tragedy such as *Julius Caesar*, in which an act of betrayal between friends tears a community apart. Once disrupted, there are two ways to

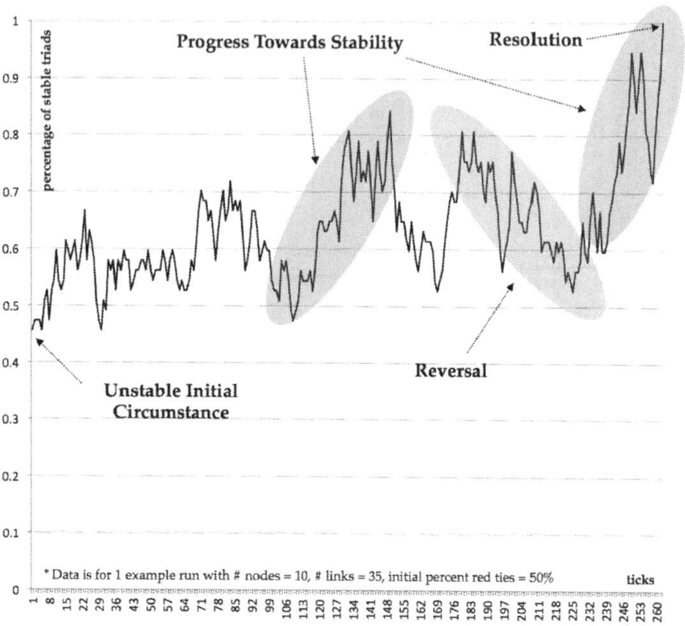

Figure 4.6 Example run showing percentage of stable triads over time.

restore stability: (1) The disturbing link can be extinguished or forced back into conformity as in *Romeo & Juliet.* (2) The old social structure can be completely unraveled and a new equilibrium established, as in *Julius Caesar.* The latter case is pictured in Fig. 4.7. At $t = 0$, the network consists of only friendship ties except for one enmity link. Universal harmony unravels in the first phase of the run—what we might call "act one"—evidenced by the rapid fall in friendship links and rise in enmity links. This phase ends when the percentage of friendship and enmity ties is equalized, indicating the original social structure has completely collapsed. A middle phase of instability and flux follows. Finally, a new equilibrium structure begins to form consisting of two polarized factions. The model terminates when social mitosis is complete. The sequence of events is reminiscent of Harrison White's assertion that "a network can be traced as similar stories appear across a spread of dyads" [25]. The "story" that spreads in this case is one of mounting enmity—an initial act of betrayal ripples outward, cleaving the network.

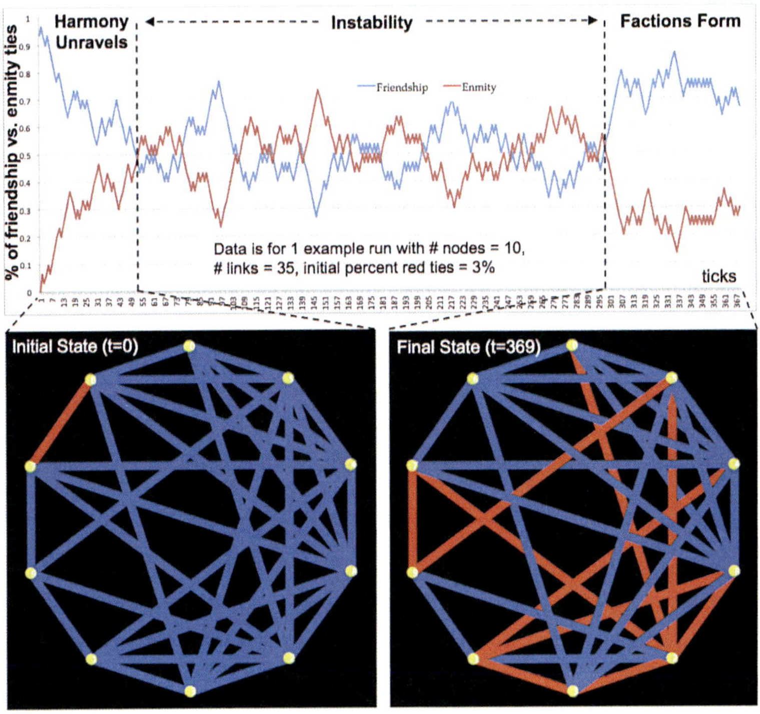

Figure 4.7 (a) Example run showing percentage of friendship (blue) vs. enmity (red) ties over time. (b) Initial and final SBT network.

Figure 4.8 Story ending, represented by stability outcome, is affected by network topology and percentage of friendship versus enmity ties.

While the model will eventually find an ending represented by a stable outcome, *when and what type of ending are indeterminate.* There are three possibilities for an incomplete graph: (i) *universal harmony*, or what we might consider a "happy ending"; (2) *polarized factions*, or what we might consider an "unhappy ending"; or (3) *mixed outcome.* The indeterminacy of the ending provides a rudimentary version of *narrative suspense.* Figure 4.8 shows the effect of different input parameters on the length and ending of the proto-narrative.

4.4 Conclusion

Lest the "proto-narrative" produced by this structural balance simulation strike us as overly simplistic, it is worth observing that the acts of betraying and befriending and the reconfiguration of social allegiances are the core events of many classical and contemporary social dramas, ranging from French court novels such as *Le Princesse de Cleves* and *Les Liaisons Dangerouse* to contemporary soap operas such as *Gossip Girl.* In a recent paper entitled "Facebook for Vikings," folklorist Timothy Tangherlini argues that the plot structures of Scandinavian story cycles can be understood in terms of shifting alliances and enmities consistent with SBT:

> In a great deal of saga scholarship there is an understandable emphasis on understanding enmity, with friendship acting as a powerful counter force (Byock 1982; Miller 1983 and 1990). Network analysis allows one to consider friendly interactions and antagonistic relationships both as individual features of the saga narrative and in concert with each other.... Perhaps one of the most complicated aspects of social interaction considered in the sagas is the selection of friends and its inverse, the selection of enemies [23].

Tangherlini analyzes the famous "Höfuðlausn" or head ransom episode in *Egil's Saga* in terms of interdependent triads between four characters—Egil, Eirik, Arinbjorn, and Gunnhild. All friendship and enmity ties are determined prior to the episode except for the relationship between Arinbjorn and Gunnhild. Tangherlini argues that the dramatic arc of the head ransom episode consists in the determination of the Arinbjorn–Gunnhild relationship and the reconfiguring of the Egil–Eirik relationship based on the stability requirements of SBT.[h] Tangherlini's approach is confirmed by a recent study of "mythological networks," in which MacCarron and Kenna find that several ancient and medieval epics, including *Beowulf* and *The Iliad*, obey structural balance rules [16].

Studies such as Tangherlini's demonstrate that structural balance theory has value as a descriptive model of socially complex narratives, but it has perhaps even greater potential as a *generative* mechanism that could be used for new story creation.

Finding realistic but tractable story generation mechanisms is an ongoing challenge for artificial intelligence researchers working on narrative. Most story generators rely on either (a) corpora of pre-existing stories (e.g., MEXICA [20]), or (b) story grammars (e.g., TALESPIN [18]). Researchers in this area are now attempting to expand the suite of generative mechanisms to include games as well as crowdsourcing [14]. In a recent paper, Pablo Gervas argues for the value of chess as a narrative generation mechanism:

> Chess provides a finite set of characters (pieces), a schematical representation of space (the board), and time (progressive turns), and a very restricted set of possible actions. Yet it also allows very elementary interpretations of game situations in terms of human concepts such as danger, threat, conflict, death, survival, victory or defeat, which can be seen as interesting building blocks for story construction [9].

There is a related body of work that makes use of sports game statistics to generate simple narratives akin to newspaper articles

---

[h]Tangherlini's study also provides a useful set of benchmarks for calibrating the SBT model. In *Egil's Saga*, Tangherlini counts a total of 316 friendship interactions and 158 enmity interactions, of which 61 are "lethal" and 97 are "non-lethal" (a roughly 2:1 ratio of friendship to enmity events). This total of 474 interactions is spread across 200 "saga actants." The ratio of events to characters that Tangherlini finds is low compared to that generated by the SBT model I have outlined. This is likely due to several factors. First, the graph for *Egil's Saga* is highly incomplete, with very low graph density. As shown in Fig. 4.4, the length of the "event history" is very sensitive to the number of links in the network: relatively complete graphs take exponentially longer to resolve than relatively incomplete graphs. Second, and perhaps more importantly, there is a key distinction that should be drawn between story-events and discourse-events. Our model generates story-events, not all of which need ultimately be rendered narratable in the final text. The events in the discourse of *Egil's Saga* have been selected from underlying story-events that are deemed worthy of narration because, for example, they involve specific characters of interest, such as Egil.

[1]. Like chess and sports, dynamic network models such as structural balance could provide a narrative generation mechanism to complement corpus analysis. They possess the added advantage of supporting social behaviors that are more complex than direct competition or physical conflict.

Most promisingly, there is growing emphasis in the interactive narrative community on story-worlds that incorporate complex and nuanced social dynamics. Notable examples include *The Sims* and Michael Mateas' experimental game *Façade* [15]. UC Santa Cruz's *Prom Week*—in which players manipulate the social interactions between high schoolers in the week leading up to prom—is perhaps the most sophisticated example of an emerging category of "social games" that incorporate sociological and psychological models into AI to "gamify" social experience. *Prom Week* makes explicit use of social networks and claims to have more than five thousand social rules guiding character behavior, including "if someone is mean to me, then someone else does something mean to them, I'm more likely to want to date that person [17]." This enemy-of-my-enemy-is-my-friend rule could be integrated easily with structural balance dynamics. As "social games" gain traction in interactive narrative, generative network models such as the one I have outlined on structural balance will have increasing value.

Acknowledgments

This article was developed through my participation in several workshops and institutes pertinent to the topics of simulation and narrative, including the NEH Institute on Computer Simulation in the Humanities (UNC, 2010), the Santa Fe Institute's Complex Systems Summer School (Santa Fe, 2012), Intelligent Narrative Technologies (Stanford University, 2012), and Computational Models of Narrative (Universität Hamburg, 2014). I also would like to thank the following individuals for reading, conversing about, or otherwise inspiring the project: Patrick Grim, Mark Finlayson, Pablo Gervas, Jan Christoph Meister, Bernhard Fisseni, Benedikt Löwe, Jason McKenzie Alexander, Aaron Bramson, Daniel Singer, Zev Berger, Marvin Croy, Paul Youngman, Mirsad Hirzikadic, Ted Carmichael, Tony Beavers, Tim Tangherlini, Tom Carter, Simon

DeDeo, Nicholas Dames, David Elson, Franco Moretti, Ryan Heuser, and Carissa Flocken.

References

1. Nicholas D. Allen, John R. Templon, Patrick Summerhays McNally, Larry Birnbaum, and Kristian Hammond. StatsMonkey: A data-driven sports narrative writer. In Mark Alan Finlayson, editor, *Computational Models of Narrative: Papers from the 2010 AAAI Fall Symposium*, number FS-10-04 in AAAI Technical Reports, pp. 2–3. AAAI Press, 2010.
2. Robert Axtell and Joshua Epstein. *Growing Artificial Societies*. The Brookings Institution, 1996.
3. Ruth Aylett. Narrative in virtual environments: towards emergent narrative. In M. Mateas and P. Sengers, editors, *Narrative Intelligence: Papers from the AAAI Fall Symposium*, number FS-99-01 in AAAI Technical Reports, pp. 83–86, 1999.
4. Dorwin Cartwright and Frank Harary. Structural balance: A generalization of Heider's theory. *Psychological Review*, **63**(5), 1956.
5. Miguel de Cervantes. *Don Quixote*. Project Gutenberg Ebook, http://www.gutenberg.org/cache/epub/996/pg996.html. July 27, 2004.
6. Charles Dickens. *The Pickwick Papers*. Project Gutenberg Ebook, http://www.gutenberg.org/files/580/580-h/580-h.htm. April 22, 2009.
7. David K. Elson, Nichoas Dames, and Kathleen McKeown. Extracting social networks from literary fiction. In Jan Hajič, Sandra Carberry, Stephen Clark, and Joakim Nivre, editors, *Proceedings of the 48th Annual Meeting of the Association for Computational Linguistics*, Stroudsburg, PA. Association for Computational Linguistics, pp. 138–147, 2010.
8. Francis Ferguson. *Aristotle's Poetics*. Hill and Wang, New York, 1961.
9. Pablo Gervás. From fleece of fact to narrative yarns. In Mark Alan Finlayson, editor, *Proceedings of Computational Models of Narrative 2012*, Ýstanbul, pp. 125–134, 2012.
10. René Girard. *Deceit, Desire, and the Novel: Self and Other in Literary Structure*. Johns Hopkins Press, Baltimore, 1961.
11. Fritz Heider. Attitudes and cognitive organization. *Journal of Psychology*, **21**, pp. 107–112, 1946.
12. Norman P. Hummon and Patrick Doreian. Some dynamics of social balance processes. *Social Networks*, **25**, pp. 17–49, 2003.

13. K. Kulakowski, P. Gawronski, and P. Gronek. The Heider balance: a continuous approach. *International Journal of Modern Physics C*, **16**, pp. 707–716, 2005.
14. Boyang Li, Stephen Lee-Urban, Darren Scott Appling, and Mark O. Riedl. Automatically learning to tell stories about social situations from the crowd. In Mark Alan Finlayson, editor, *Proceedings of Computational Models of Narrative 2012*, Ýstanbul, pp. 142–151, 2012.
15. Michael Mateas and Andrew Stern. Façade: An experiment in building a fully-realized interactive drama. Presented at the Game Developers Conference, Game Design track, San Jose, California, 2003.
16. Pádraig MacCarron and Ralph Kenna. Universal properties of mythological networks. *EuroPhysics Letters*, **99**, p. 28002, 2012.
17. Josh McCoy, Mike Treanor, Ben Samuel, Brandon Tearse, Michael Mateas, and Noah Wardrip-Fruin. Authoring game-based interactive narrative using social games and Comme il Faut. Presented at the 4th International Conference & Festival of the Electronic Literature Organization, 2010.
18. J. Meehan. Tale-spin. *Proceedings of IJCAI*, pp. 91–98, 1977.
19. Franco Moretti. Network theory, plot analysis. *New Left Review*, **68**, pp. 80–102, 2011.
20. Rafael Pérez y Pérez and Mike Sharples. MEXICA. *Journal of Experimental and Theoretical Artificial Intelligence*, **13**, pp. 121–139, 2006.
21. Graham Sack. Quantifying imaginary social forms: Character networks in the 19th century British novel. Master's thesis, Columbia University, 2011.
22. Sci2 Team. Science of Science (Sci2) Tool (Indiana University and SciTech Strategies, http://sci2.cns.iu.edu). 2009.
23. Timothy Tangherlini. Facebook for Vikings: Social network analysis and Egils Saga. Presented at the Society for the Advancement of Scandinavian Study Conference, 2011.
24. Zhigang Wang and Warren Thorngate. Sentiment and social mitosis: Implications of Heider's Balance Theory. *Journal of Artificial Societies and Social Simulation*, **6**(3), 2003.
25. Harrison White. *Identity and Control: How Social Formations Emerge*. Princeton University Press, New Jersey, 2008.

Chapter 5

Modeling the Communications Circuit: An Agent-Based Approach to Reading in "N-Dimensions"

Jeremy Throne

Literature Department, University of California, Santa Cruz, 1156 High Street, Santa Cruz, CA 95064, USA

jthrone@ucsc.edu

Textual critic Jerome McGann's call for reading in "n-dimensions" invites a move away from viewing texts as static forms produced by highly orchestrated attention from authorial, editorial, social, or economic decision makers and toward more dynamic models of texts as outcomes produced by numerous uncoordinated local encounters. One of the more extensive efforts to describe these momentary encounters resides in the field of book history, where scholars often call upon Robert Darnton's model of the communications circuit in order to map the routes a text travels as it circulates among authors, publishers, printers, shippers, sellers, and readers. As it has thus far been developed, the communications circuit functions primarily as a descriptive tool. However, when approached from the perspective of agent-based modeling, this static form leads us toward a host of potential alternative histories

Complexity and the Human Experience: Modeling Complexity in the Humanities and Social Sciences
Edited by Paul A. Youngman and Mirsad Hadzikadic

ISBN 978-981-4463-26-3 (Hardcover), 978-981-4463-27-0 (eBook)
www.panstanford.com

contained in and excluded by the historical record. The second half of the paper outlines an agent-based model that reveals a two-tiered system embedded in Darnton's paper model of the communications circuit: The result suggests scholars seeking to base their research upon the circuit or a similar model may want to describe, in some fashion, how they see responsibility for the circulation of a text distributed among the components of the circuit, as well as how they see these components interacting.

5.1 Introduction

> Every text, every element of every text, is n-dimensional, depending on what you choose to regard as contextually relevant.
>
> —Jerome McGann, "Electronic Archives and Critical Editing"[1]

Textual critic Jerome McGann frequently describes texts as "n-dimensional." His use of the term is an explicit attempt to build bridges between the fields of complex systems research and literary studies. He makes this clear while reflecting on his work with the digital reading environment IVANHOE, which began as a collaboration between McGann and Johanna Drucker in 2000. "The formulation 'n-dimensional,'" McGann argues, "is important as a rhetorical move to situate IVANHOE within a context where quantum mechanics and self-organizing systems can meet the hermeneutical tradition's commitment to the search for meaning"[2]. McGann's interest in and commitment to multidimensional approaches to meaning is a fundamental part of his approach to editorial theory.

A brief sense of McGann's philosophy can be obtained by juxtaposing the views of Robert Hirst, general editor of the Mark Twain Project, and Kenneth Price and Ed Folsom, co-directors of the Walt Whitman Archive. Mark Twain, as you may know, frequently expressed displeasure with the way his texts were edited.[a] Responding to this sentiment, Hirst claims, "the Mark Twain Project editors are therefore interested in giving Mark Twain the text he wanted but never got"[3]. Meanwhile, at the Whitman Archive, Price and Folsom argue, "in the last analysis, we are not editing Whitman for Whitman but for ourselves and for

[a]For a selection of Twain's comments on editors and editing, see Barbara Schmidt's popular Twainiana Web site at http://www.twainquotes.com/Editors.html.

all those interested in him in our historical moment"[4]. Hirst's position is a contemporary instance of a line of editorial theory established in the United States in the 1960s by bibliographer Fredson Bowers. Bowers saw editing as an attempt to purify and preserve an author's final intentions for a text [5]. McGann—particularly in his 1983 work *A Critique of Modern Textual Criticism* —rejects the vision of editing offered by Bowers. Denying the supremacy of authorial intent, McGann offers a model of texts as mutable products that are repeatedly fashioned and refashioned to reflect the socio-historical contexts that surround them at any given moment in time. McGann's advocacy for what he calls the "socialization of texts" provides a foundation for the view endorsed by Price and Folsom; his interest in asserting the n-dimensional text is an extension of this position: it is the "socialization of texts" to the nth degree: a call to study a text not only in its moment, or in our moment, but in all moments—some vouched for by our historical records and some not. Reading in n-dimensions, then, is not simply about illuminating and/or explicating various aspects of a text, it is a way of asserting that we are implicated in and bound up with the texts around us. The idea that readers and texts are one and the same is a central part of IVANHOE. Unfortunately, the project is currently inaccessible—it was taken off-line in January of 2012—however, the bridges it built remain; what follows is an attempt to employ agent-based modeling to test the span connecting complex systems research and literary studies.[b]

I begin by reading the adjective "n-dimensional" as an invitation to treat texts not as the "well wrought urns" of New Critical attention, nor as alternatingly over and under wrought sites of deconstructive play, but as emergent phenomena. In other words, the phrase "n-dimensional text" invites us to move away from a view of texts as static forms produced by highly orchestrated attention from authorial, editorial, social, or economic decision makers and toward more dynamic models of texts as outcomes produced by numerous uncoordinated local encounters.

One challenge to enacting this shift lies in describing these momentary encounters in ways that acknowledge and unwind the

[b]For the current status of IVANHOE, see http://www.ivanhoegame.org. The 2013-2014 Praxis Program fellows at the Scholars' Lab at the University of Virginia are working to redesign the game as a WordPress theme. For more on their project, see http://www.scholarslab.org/tag/ivanhoe/

n-dimensions that constitute any particular text without giving in to the temptation to erect hierarchal structures where they may not exist. From the perspective of agent-based modeling, this challenge takes the form of a deceptively simple question: How do we parameterize a book? One of the more extensive efforts to answer this question currently resides in the field of book history.

5.2 Book History

Book history is an interdisciplinary effort to study texts in their literary, cultural, and material contexts that begins to emerge as an institutional presence in the United States in the early 1990s with the founding of the Society for the History of Authorship, Reading, and Publishing (SHARP) and the launch of Book History, the flagship journal for the field, later in the same decade.

Book historian Michael Winship's views are representative of the dominant perspective in the field: "authors do not make books." Winship goes on to argue "the printed book in particular is the product of the collaborative efforts of many different people: laborers and craftsmen, publishers and editors, businessmen and patrons" [6]. His claims sketch the beginnings of a move from static to dynamic models of texts: Like McGann, Winship rejects the supremacy of the author. This move creates space for exploring some of the less recognized factors involved in literary production, but it stops short of taking a position on whether interactions among these factors produce one or many texts. McGann's figure of the n-dimensional text, the one text that is many, can be read as an attempt to go a step further. The n-dimensional text harnesses but does not resolve the tension between a desire to connect texts and a desire to treat each as a discrete entity. However, the n-dimensional text has not yet reached the mainstream, even among book historians. Its presence is scarce, for example, in many of the important volumes recently compiled to survey the field.

Readers of *The Book History Reader*, *An Introduction to Book History*, and *A Companion to the History of the Book* are likely to read of McGann's push for the "socialization of texts," but will find n-dimensional texts rarely mentioned. However, readers of these volumes will not fail to be introduced to the communications circuit, one of the more established attempts among book historians to conceptualize the kind of multifaceted perspective championed by Winship.

5.2.1 The Communications Circuit

The communications circuit is an attempt to map the routes a text travels as it circulates among authors, publishers, printers, shippers, sellers, and readers. It was proposed by Robert Darnton in the early 1980s in an attempt to quell the rise of what he characterized as "interdisciplinarity run riot." Book history, he worried, "looks less like a field than a tropical rainforest"[7]. Darnton's occasionally playful diagnosis of the field's ailments has become a classic article among book historians and the appeal of his model remains strong. Nearly 30 years after its proposal, the communications circuit has become a standard feature of course syllabi and other introductory materials for the field, while retaining a role in some of the more cutting edge book history scholarship.

The influence of Darnton's attempt to track the circulation of texts is evident, for example, in the work of Fiona Black, whose advocacy for a GIS-inspired approach to book history is one of many efforts to infuse the field with digital technology. Included among the articles found in the first issue of the journal *Book History*, "Geographic Information Systems: A New Model for Book History," by Black and her colleagues Bertrum MacDonald and J. Malcolm Black, transforms the stops on the communications circuit into layers of data on a virtual map. Black's work, and the work of others like her, represents an important advance in our thinking about books, their lives, and their afterlives. It is a move from pen and paper sketches to full-color representations of texts and the worlds in which they move. But perhaps even more importantly, it points the way toward the pressing work yet to be done.

5.3 The Communications Circuit: An Agent-Based (RE) Model

As it has thus far been developed, the communications circuit functions primarily as a descriptive tool. Even in the digital form championed by Black and her colleagues it remains an essentially static model of literary circulation. However, when approached from the perspective of agent-based modeling, this static form leads us toward a host of potential alternative histories contained in and excluded by the historical record. The recognition of these alternatives is a preliminary step toward the creation of what we might call a textured or textural history: A history that takes the smooth polish of official history and pairs it with the pockmarks that often appear when that history is seen from another angle.

Illuminating the textures on our historical records—the polish of official history and the pockmarks of that history when seen from below—and bringing them together to create what I call textural histories is one of the more exciting opportunities that may result from the union of complex systems and literary studies sought by McGann and other members of what Johanna Drucker has dubbed the "lunatic core" of the digital humanities community [8].

One way in which agent-based modeling may help with this process is by taking up the communications circuit and transforming it from a directive designed to quell interpretative chaos into a tool for investigating multiple possible interpretations of the past. Modeling the communications circuit allows us to read not simply against the grain of history, to invoke Benjamin's well known dictum, but to accentuate the grain of history. This shift may be necessary because, as any woodworker can tell you, going against the grain eventually produces a smoothness of its own. In what remains of this paper I will outline an approach I am developing that draws upon agent-based modeling to accentuate the grain in Darnton's communications circuit.

Each of Darnton's six roles becomes an agent in my model:

- Authors, as I have coded them, simply want to publish books. Each author begins with a single manuscript and wanders across the uniform grid of squares that constitutes the environment, randomly moving one square per time step (or "tick") of the model, in search of a publisher. If an author meets a publisher, the author leaves the manuscript with that publisher and then continues wandering the environment.
- Publishers in the model are seeded at random throughout the environment. They accept manuscripts from authors and transfer them to printers.

- Printers receive manuscripts one at a time from publishers. They hold onto their work until a shipper arrives.
- Shippers, like authors, wander the environment randomly, one patch per tick. If a shipper meets a printer with jobs waiting, the shipper takes one job from the printer and sets out in search of a bookseller. While searching for a bookseller, the shipper will continue to take on an additional unit of cargo each time a printer with jobs waiting to ship is encountered.
- Booksellers are seeded at random throughout the environment. When a shipper encounters a bookseller, all of the cargo the shipper is carrying is transferred into the inventory of the bookseller.
- Readers in the model wander, randomly, one patch per tick, through the environment. When a reader encounters a bookseller, the reader acquires a book and the inventory of the bookseller is depleted accordingly. In addition, readers have a second function: If a reader encounters an author while wandering the environment, the reader encourages the author to produce another book. When the author receives enough encouragement, another manuscript is produced, beginning the cycle again as the author sets out in search of a publisher for the new manuscript.

5.4 Limitations

The simplified view of literary circulation embodied in the model is both a limitation and an asset.

The desire for distribution that drives authors in the model to seek out publishers for their manuscripts excludes from the category of author those who do not seek to place their work in the public sphere. Diary writers, for example, do not qualify as authors in the logic of the model as I have currently explained it. However, one might expand the model to incorporate these important producers of literary texts—perhaps by adding a new type of author-agent to the environment, perhaps by explicitly altering our understanding of authorship to incorporate work brought to the attention of publishers by someone other than an author. Consider, for example, the work of Anne Bradstreet, whose manuscripts were famously published in London by her brother-in-law, John Woodbridge, while she remained in the New World. Is Bradstreet an author? Coding her behavior stages the tension between creator and conveyor that underlies discussions of authorship. This tension is elided in more traditional models of the communications circuit; it lies dormant, for example, in Darnton's paper model, but is awakened upon the transference of his work to the digital realm.

Publishers are flattened in my current model. The flattening stands in contrast to the large and colorful body of scholarship devoted to unearthing details about the lives of publishers. Integrating this material is one of the challenges awaiting future modelers. Prestige, for example, is one trait that many scholars of the publishing industry take note of in their studies. Winship, for example, in his study of the nineteenth-century publishing house Ticknor and Fields, notes that it was the most esteemed publisher of its day. Distinctions such as this may be coded into models to draw distinctions between publishers and increase the diversity of the publishing world realized via the model. Such additions then require the addition of an ability to recognize prestige, but to whom should this ability be assigned and to what degree? Deciding these issues emphasizes a need not only to describe the qualities of a publishing house, but also to dwell upon how those qualities may have been received across various subsets of society including, authors, printers, shippers, sellers, and readers, as well as other publishers.

Printers, as I assume in my model, are known to publishers and receive business directly from them. They do not search for manuscripts, but rather wait for them to arrive. Printers also share the same working conditions: They may only work on one manuscript at a time; and, they must complete one job before taking on another.

The shippers I have created are nomadic, moving from place to place acquiring and depositing cargo as space allows. They are neither called for by printers, nor expected by sellers. Such behavior provides a baseline against which one can explore how the establishment of more conventional shipping and receiving routes may impact literary circulation. Frank Mott, for example, in his multivolume history of magazine publishing in the U.S. argues that it was the development of standard distribution channels—roads, post offices—that allowed the popularity of magazines to grow. By controlling the movement of shippers in more or less standard ways one might test this claim.

The homogenous nature of these businesses is admittedly unrealistic, as is the homogenous nature of the behavior of agents at other points in the model. Sellers, for example, make no judgments regarding their inventory in the model I have constructed; if a shipper arrives with books, they accept them. Readers are assigned two simple traits: They acquire books when they encounter them, an act that may be thought of as reading, purchasing, or borrowing; and, they praise authors when they encounter them. Interactions between readers and authors Contact with readers currently spurs authors to produce more books; it could also be used to discourage authors, but I have not yet incorporated any form of negative feedback into the model. Readers also read insatiably in the current model, another trait that may be altered to further diversify the environment. In constructing the behavior of a reader, the difficulty of determining how readers respond to texts that plagues reader response theory is juxtaposed with an ability to determine how a range of reader responses may impact the circulation of a text.

A chief benefit of attempting to model these behaviors is that one may then alter them to determine how preferences—for certain publishers, printers, or shippers; or for maintaining particular inventory levels—may impact the circulation patterns. What appear, on one level, to be simple caricatures, offer, on another level, the benefit of observing how the model as a whole may function in a uniform environment. Simplification also creates space for observing how altering the behavior of an agent or group of agents may impact the outcomes produced by the model as a whole.

Thinking through possible ways to address these issues, striking a balance between fidelity to the real world and forays into abstraction, helps to illuminate ways in which book history, sometimes framed as an escape from theory and a turn toward empirical results, can also become a deep engagement with fundamental questions of literary criticism. It also highlights how fundamental questions of literary theory—What is an author? What role do readers play in the circulation of works of literature? How do material circumstances impact literary circulation?—bring attention to the same balance of real and abstract that must be negotiated as part of the model-making process.

5.5 Applications

Despite its flaws, even this simple model has something to offer book history research. From a theoretical perspective, the model stands as an invitation to what McGann calls "quantum" reading, the practice of reading competing versions of the literary together without attempting to impose an order upon them.[c] The model encourages this process by producing a different version of literary history each time it is run. Finding variety through a set of parameters draws attention to the multiple ways in which visions of the past may unfold.

From a more practical point of view, modeling the communications circuit draws attention to the gaps and grounds of literary history by making explicit what is included and excluded from any given perspective on the past. The act of coding also enables us to consider the effect of altering these inclusions and exclusions. In a similar vein, adopting a model-based approach to literary history makes one acutely aware of both the written historical record and its more speculative (but no less necessary) alternatives. This combination of factual and contingent knowledge provides a key component in the transformation from textual to what I am calling textured versions of literary history. An ABM-based approach aids this transformation by simultaneously enacting what we know and showing us how contingent that knowledge is.

Modeling the communications circuit provides an opportunity to observe textural history in action. Darnton's diagram is informative and has proven itself to be a cornerstone for many textual histories; it is also a static, flattening approach to the past. However, in a virtual form it provides a dynamic tool for literary exploration.

[c]For an example of quantum reading, see the essay "Mr. James and His Discovery" in McGann's *The Scholar's Art: Literary Studies in a Managed World*. Chicago: University of Chicago Press, 2006.

5.6 A Model in Motion: Demonstration and Results

One interest of mine concerns the features that make up the circuit. Multiple alternatives to Darnton's model have been suggested, each with its own set of components. For example, Darnton himself, when asked in 2007 to reflect on the model, suggests adding emphasis on the activities of smugglers, literary agents, and traders, among other factors [9]. Armed with a digital model we can begin to explore the impact of these extensions. At what point does smuggling become a significant enterprise with power to alter the circulation of a text? How do the activities of promoters change distribution patterns? Questions like these are likely candidates for future models. For the moment, I am content to explore how the composition of a communications circuit as Darnton originally envisioned it influences the circulation of literary texts. To answer this question, I tested my model by creating an environment populated with 5 authors and 100 readers. I then introduced into this environment varying numbers of publishers, printers, shippers, or sellers.

My exploration suggests that for the given population of authors and readers, movement of texts through the circuit appears to be more closely connected to the presence of printers and shippers than it is to the number of publishers or sellers in the environment.

I began my exploration by checking to see how long it would take texts to reach 10% of the readers in the environment when various numbers of intermediate agents were present. For example, when the model was run with 5 authors, 1 publisher, 1 printer, 1 shipper, 1 seller, and 100 readers, it took roughly 2 ticks for texts to reach 10% of the readers in the environment in my first trial. I then repeated the investigation holding all parameters constant except for the number of publishers, which was raised to 10 publishers, then 100, and 1000. I conducted this set of experiments 100 times for each number of agents. Figure 5.1 summarizes the results of my exploration.

Letting the model run to reach 50% of readers also shows a decrease in time when the number of printers or shippers is increased. As with the tests conducted to reach 10% of readers, changing the number of publishers or the number of sellers in the environment appears to have little impact. Figure 5.2 summaries results for these tests.

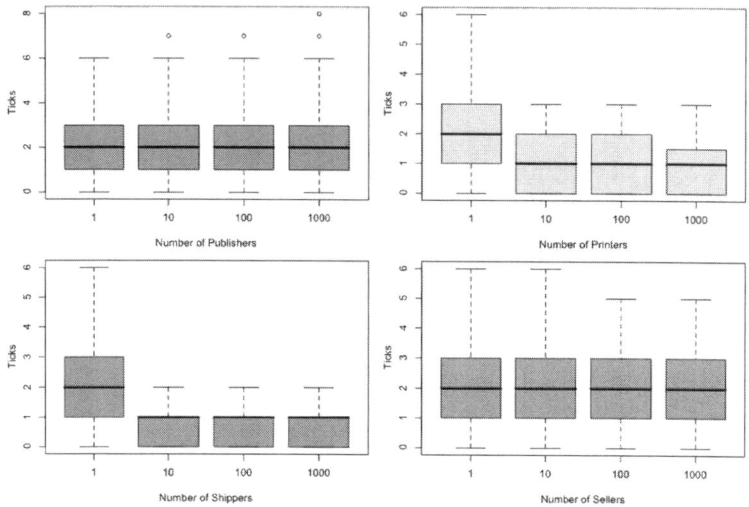

Figure 5.1 A comparison of time to reach 10% of readers for each type of agent (100 runs per number).

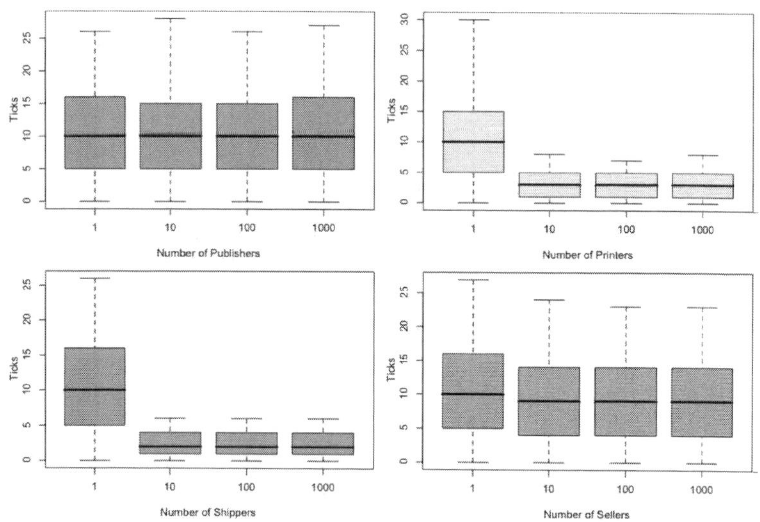

Figure 5.2 A comparison of time to reach 50% of readers for each type of agent (100 runs per number).

I obtained similar results by letting the model run to reach 90% of readers. I have not yet tested to determine when these outcomes first appear, but it seems the activities of printers and shippers begin to distinguish themselves from those of publishers and sellers fairly early on. I did, however, conduct further tests on the 1–10 agent range for all four types of agents. For these tests I decreased the size of the environment by half in order to decrease the runtime for the exploration and allowed the model to run until texts had reached 90% of readers in the environment. Figure 5.3 summarizes the results

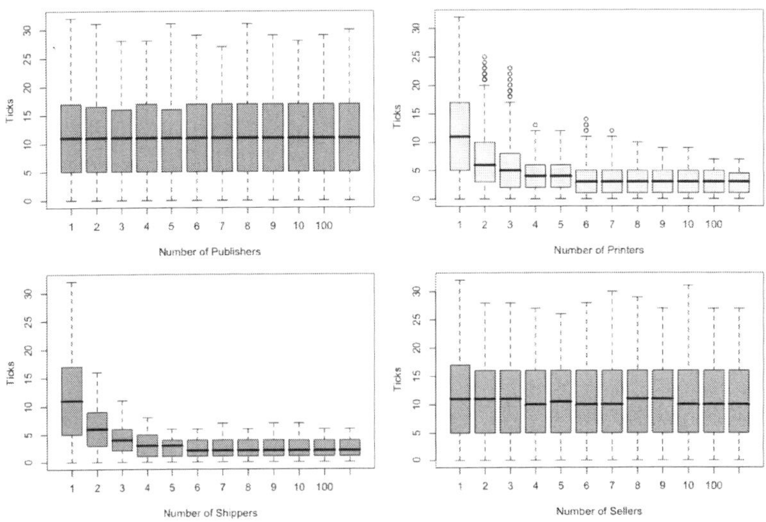

Figure 5.3 A comparison of time to reach 90% of readers for each type of agent (100 runs per number).

The 90% tests stress that the median time it takes for texts to reach readers appears to hold relatively constant no matter how many publishers or sellers are introduced into the environment. A circuit with 1 publisher has a median time near 10 ticks, as does a circuit with 1000 publishers. Increasing the number of printers or shippers, however, produces a noticeable decrease in the number of ticks needed to reach 90% of readers, with the biggest drop occurring in the jump from 1 to 2 agents and median times stabilizing from six agents on.

Two types of responses appear to have emerged from the tests I have conducted: Increasing the number of printers or shippers reduces the time it takes for texts to reach readers, while the result of increasing the number of publishers or sellers is less clear.

5.7 Conclusions

I offer this work as an example of how agent-based modeling may be used to accentuate the grain or bring out the latent textures inherent in a static model of the past such as Darnton's communications circuit. I take my results as a provisional, rather than a definitive statement about how the communications circuit operates. I look forward to exploring this in the future and welcome comments from all interested parties. It is not the results themselves I see as significant, but the fact that there are results. The digital model I have constructed reveals an apparently two-tiered system embedded in Darnton's paper model of the communications circuit: Printers and shippers exert more influence over how texts flow through the circuit than publishers and sellers. This result strikes me as significant for a couple of reasons: First, it draws attention to elements of literary circulation that are commonly thought of as middlemen, rather than key players in literary history. The model suggests that it may perhaps be the activities of these middlemen that are in fact key to any understanding of how texts move through an environment and may, with further testing, provide a guide to research in areas like bestsellers and other studies concerned with questions of distribution, or lack of distribution. Second, and perhaps, more importantly, the results suggest that one should not assume textual circulation to be equally distributed across the components of a communications circuit and that, at the very least, scholars seeking to base their research upon the circuit or a similar model may want to describe, in some fashion, how they see responsibility for the circulation of a text distributed among the components of the circuit, as well as how they see these components interacting. Agent-based modeling provides, to my mind, a likely tool for exploring these questions.

References

1. McGann, J. (2010). Electronic Archives and Critical Editing, *Literature Compass*, 7.2, 37–42.
2. McGann, J. (2003). Texts in N-Dimensions and Interpretation in a New Key (Discourse and Interpretation in N-Dimensions), *TEXT Technology*, 12.2, 1–18.
3. Hirst, R. (2007). Textual Editing at the Mark Twain Project: A Brief Account, *Mark Twain Project Online* (Internet), available from http://www.marktwainproject.org/about_hirst_essay.shtml.
4. Folsom, E., and Price K. (2005). *Re-Scripting Walt Whitman: An Introduction to His Life and Work* (Internet), Blackwell, Oxford, UK, available from http://www.whitmanarchive.org/criticism/current/anc.00152.html.
5. Bowers, F. (1964). Some Principles for Scholarly Editions of Nineteenth-Century American Authors, *Studies in Bibliography*, **17**, 223–228.
6. Winship, M. (1995). *American Literary Publishing in the Mid-Nineteenth Century: The Business of Ticknor and Fields*, Cambridge UP, Cambridge, UK, p. 188.
7. Darnton, R. (1982). What is the History of Books?, *Daedalus*, 111.3, 65–83.
8. Nowviskie, B. (2009 June 6). sketching Ivanhoe, nowviskie.org [Internet], available from http://nowviskie.org/2009/sketching-ivanhoe/.
9. Darnton, R. (2007). "What is the History of Books?" Revisited, *Modern Intellectual History*, 4.3, 495–508.

Chapter 6

The Potential of Agent-Based Modelling for Historical Research

Marten Düring

University of North Carolina at Chapel Hill,
551 Hamilton Hall, CB# 3195, Chapel Hill, NC 27599, USA

martenduering@gmail.com

Agent-based modelling can be a useful aid to historical explanations, in the sense that it can help test hypotheses and yield new insights through the process of model-building. All it takes are a limited number of causal factors for the analysis, good source availability, stable conditions and well-defined hypotheses. While today modelling is more common in Archaeology than in any other historical discipline, the case studies show that there are no limitations regarding the time spans considered or the historical data used. Agent-based modelling and other computational methods in the historical disciplines bear the potential to be combined with counterfactual history in that both seek to open up opportunities to study historical events that could have happened and thus help scholars sharpen their analyses.

Complexity and the Human Experience: Modeling Complexity in the Humanities and Social Sciences
Edited by Paul A. Youngman and Mirsad Hadzikadic

ISBN 978-981-4463-26-3 (Hardcover), 978-981-4463-27-0 (eBook)
www.panstanford.com

6.1 Introduction

As agent-based models slowly become more common in the humanities, their application in the historical and archaeological disciplines increases as well. This paper investigates the current practice of agent-based modelling in the historical disciplines, presents a selection of models, discusses their goals and links them to the still marginal practice of counterfactual thought experiments in history.

Historical research is commonly understood as the art of identifying, interpreting and contextualizing facts from the past, followed by the construction of narratives with the aim of explaining past events [20, p. 7]. Historian Johann Gustav Droysen sums up these efforts as the triad of source-criticism, heuristics and interpretation. Historians excel at acknowledging ambiguities and contradictions, and at taking into consideration historical context. The results of their work, typically narrative based, are constructions based on a fallible interpretation of typically highly fragmentary sources.

Agent-based models, however, require the notoriously difficult yet also arguably very thought-provoking process of simplification of the complexities and ambiguities of historical events and human behaviour. This implies that any causal relations within the events studied need to be first interpreted based on often fragmentary sources and then grounded on either facts, theories or assumptions. The process of abstracting a thus highly simplified representation of an historical event is followed by its quantification. This means that concepts such as "happiness", "thirst" or "firepower" need to be quantified; which in turn also means that meaningful causal relations need to be expressed in numbers: It needs to make sense when a thirst level of say "5" triggers a specific action. While this may be easier for some concepts than for others, it remains a challenge to find appropriate approximations for every causal factor.

Computer games, especially simulations of past societies have been a point of reference for many scholars interested in agent-based simulations in the archaeological and historical disciplines. Of particular note is the didactic potential of these games [for an extensive collection of materials see reference 26]. Karsten Weber begins his article on computer simulations in history with a reference to common, round-based strategy games such as "Civilization" or "Anno 1701" which contain some basic factors and dynamics that determine the evolution of past societies [37]. These artificial societies or settlements typically start out small. A few warriors and peasants are needed to build an empire from the bottom up. There are resources such as gold, metal or timber that need to be gathered, there is division of labour between peasants who gather these resources and warriors who defend their own and attack other settlements. Progress in the development of one's settlement—for example a certain number of buildings, gathered resources or agents—opens up new options for further advancement of one's settlements. New types of buildings, weapons and agents become available to the players as they make progress in the game. This progress leads players to further develop their settlements, to successfully defend them against competitors, to explore and occupy larger parts of the game world and, sometimes, to destroy all competing settlements.

These models of societies are obviously designed to entertain and not to mimic the rise and fall of past empires: historic societies had more than a handful of social roles, depended on more than three resources, did not automatically develop new skills based on rules and could not have evolved independently from each other. There are plenty more reasons to conclude that current computer games are not adequately differentiated, but crude oversimplifications of past events. Still, these games can be understood as representations of real-world processes which recognize basic rules of social evolution: The development of new technology depends on the availability of resources and prior developments, humans and the places they live in need space and face competition from other social groups [15,25,27].

Weber goes on to propose increasing the complexity of these games so that they are able to appropriately reproduce historical processes. A "historical microscope" which mirrors historical events and allows them to be observed inside the model and "[t]o sum up the hopes for future applications of computer simulations in the historical sciences, this would mean that we had a time machine which enabled us to directly observe historical processes" [37, p. 96]. Model-builders, at least at first sight, seem to share Weber's enthusiasm:

> With ABM, we can run our artificial society through a series of artificial histories. Then, we select the runs which seem to best correspond to the actual history which did occur. We look at our parameter settings for these best runs, and if our model has been carefully designed and validated, we will know something new about the original society [11, p. 2].

They propose "A Virtual Prehistory" [17, p. 78] through models and the possibility that "the tape can be rewound" numerous times. [as cited in 24, p. 11]. Some suggest that simulations finally allow for "verifiable interpretations" instead of "unproven assertions" and that they are tools to "break these cycles of academic claim and counter-claim" [24, p. 1]. Statements like these could easily be interpreted in a sense that understands our perception of past events as something objective and not retrospectively constructed: "real" history vs. "artificial". To be clear: all of the quoted researchers are very careful to present their results in a differentiated and self-reflective manner. Despite these bold promises, the focus of most projects has been to examine highly specific selected explanatory theories that seek to explain the dynamics of historical events.

6.2 Computational Methods and Counterfactual Thinking in History

Predictions of the future can obviously not be verified in advance of the events they describe. Historical analyses on the other hand have to deal with the problem of retrospective determinism. The authors of the report of the 9/11 terrorist attack found a nice metaphor to describe this problem:

> But the path of what happened is so brightly lit that it places everything else more deeply into shadow [as cited in 31, p. 3].

In order to fully grasp the significance of individual historical facts and constructed causations, historians need not only focus on what is "brightly lit" but also on what lies in the shadows.

Hotly contested by some and defended by others, counterfactual thinking has, until now, played a minor role in history despite its support by prominent figures such as Niall Ferguson, Thomas Carlyle and Winston Churchill [32, p. 14; most recently 12]. Tetlock, Lebow and Parker, however, argue that any form of historical reasoning involves counterfactual thinking: If we for example understand event A as very significant for the outcome of a historical event, that implies that we have considered how things would have evolved differently, had A not happened [similarly: 22]. Historical analysis, then, simply cannot do without at least implicit counterfactual thought experiments in order to question and re-evaluate one's own earlier historical interpretations [31, p. 25]. Proponents of counterfactual historical research argue that its benefit lies in the rigid and open re-examination of seemingly "brightly-lit paths", i.e., uncritically accepted truths. Both counter-factual thinking and simulations seek to sharpen and question seemingly obvious interpretations of causality.

Computational methods such as agent-based modelling are—as I will demonstrate—particularly well suited to bring to our attention events that were likely to happen but did not occur. Case studies in natural language processing have identified times when strikes could have been very likely to occur [36]; methods derived from social network analysis could identify individuals who, despite their close ties with other helpers and thus a high potential to act, abstained from any active support for Jewish refugees during the Holocaust [8]. This dimension of computational methods in the historical discipline has not yet been explored sufficiently and thus needs to remain tentative at this stage. Agent-based models, however, offer great potential.

6.3 Agent-Based Modelling of Historical Facts

Given that most agent-based models of historical events were not developed by trained historians but by computer scientists, anthropologists and epidemiologists, I work with a broad definition of the terms "historical" and "archaeological" and have considered any model which matched its findings against the outcome of a specific past event [for a continuously updated bibliography see my website, reference 9].

Similar to the aforementioned computer games, agent-based models of historical processes have concentrated on agriculture, economics, cities and military action. Since most model builders were not historians by training but scholars from other disciplines who used historical data for their research, their questions and answers will match those of historians concerned with the same period only to an extent. However, similar to ancillary sciences in history, such as the study of numismatics and demography, they promise to make important contributions to the interpretation of historical processes.

In comparison to their application in the social and natural sciences, the overall number of agent-based models dealing with historical processes is rather small. Models are most often used in archaeology and prehistory, possibly due to the fact that these disciplines have traditionally kept closer ties to the natural sciences. Less work has been done on historical processes from the Medieval Ages onwards. For the purpose of this paper, I have decided to find a balance between models based on archaeological, ancient and modern history with the aim of investigating models built based on a large variety of data.

The case studies I chose for this paper cover time spans between a few hours (Battle of Trafalgar), two years (Influenza infection in Manitoba) and 700 years (settlement structures of the Anasazi). Similarly heterogeneous is the data that these models rely on. Among the sources are log books of ships and other sources from the Napoleonic Wars, death registers, notes of fur traders and the analyses of tree rings and historical climatology. There are of course other similarly suitable case studies [6,19,21,23,29].

Many case studies share excellent historical records of the respective object of study compared to similar cases. They aim to take into account all of the available knowledge and transfer some facts from similar cases. Still, some important factors are merely based on assumptions. The nature and consequences of which could have been explained in greater detail. In my discussion of the four case studies, I will not focus on the technical design of the models themselves but rather on the following questions:

(1) Which question does the model seek to answer? Why can it not be answered using "traditional" methods?
(2) How are fragmentary historical sources used to build the model?
(3) To which extent does the model consider the peculiarities of historical cases such as unpredictable events?
(4) What do researchers claim they have learned from their model?

All models neglected the effects of the actions of individuals and fully relied on the interactions of groups of agents. It remains unclear whether this was done in an effort to simplify the model or a consequence of fragmentary sources. Either way, they result in a decisive positioning of their research towards structural explanations. To my knowledge, there are no historical models to date that allow both collective and deviant individual behaviour inside the same model. To sum up, simulations have until now been used as highly specific tools that help scholars convincingly rule out some of the competing explanations.

6.4 Case Studies

6.4.1 The Disappearance of the Anasazi

The question why by ca. 1280 ad. Ancestral pueblo tribes living in Mid-Western North America had all of a sudden disappeared from the region has remained a riddle. Teams of Archaeologists, Paleoclimatologists, Agriculturists, Hydrologists and other scientists have gathered data describing the conditions for agriculture in the region. Kohler et al. have designed several computer models of the Anasazi population and the land on which they lived. Its most recent version models the availability of rain- and groundwater, estimated crop yields and consumption, quality of soil, remains of housing, and contemporary cultural practices among descendents of the ancestral pueblos. Agents in this model produced and consumed crops, moved to different parts of the land, grew families and died. Their descendents left their families, founded new families and continued to farm the land [5].

The hypothesis that changes in the climate were responsible for the population leaving the area dominated the design of the model. At the same time, no man-made sources other than the remains of the dwellings are available. The researchers conclude that despite difficult conditions for farming, at least 400 of the 700 inhabitants could have survived in the area and must have done so under similar conditions in the past. This at least partially refutes the earlier hypothesis:

> Our simulations are not mature enough to specify the exact combination of causes that led to the periodic aggregations of

> population in the Mesa Verde region or the exodus that began by 1260 and ended shortly after 1280. The computer models make it quite clear, however, that the people living in this area (and presumably elsewhere in the northern Southwest) faced troubles on any fronts in the 13th century. The negative trends probably multiplied as time went on; for example, increased deforestation most likely limited the availability of game and some important nondomesticated plant foods, such as piñon seeds. These declines in turn fostered increased dependence on maize and its exchange and on domesticated turkey, which could be fed maize. If the exchange of maize, perhaps in the context of feasting, was also a chief lubricant for social relationships, then both subsistence and society were precariously poised on the uncertain fortunes of farming.

According to the researchers, the deteriorating conditions of life in the Mesa Verde valley can partly be explained by the actions of its inhabitants [17]. Simulations were used as one of many approaches in this research and correspond closely with the findings of other analyses [most recently: 18]. The available reports do not always make it clear which finding is to be traced to which method, it seems, however, that the argument for the anthropogenic causes of decreased availability of resources can be traced back to the simulation. Kohler et al. developed the model over several years, continually improved it and interpreted the results in close consideration with the existing historical record to ensure validity.

6.4.2 The Battle of Trafalgar

On 21 October 1805, the British Admiral Lord Nelson set out to attack the French and Spanish Navy on the southwest coast of Spain. Nelson had 27 ships under his command while his adversaries had a total of 33. Despite this imbalance, the French lost 22 of their ships; the British Navy lost none. It is widely held that Nelson's deviation from common naval strategy led to his victory: Instead of forming a line of his ships like his enemies did, Nelson chose to arrange them in to columns approaching their adversaries in a perpendicular angle. F. W. Lanchester developed a set of differential equations which he used to prove this point. According to Lanchester, Nelson's best guess must have been to split the enemy line in smaller fractions.

Giuseppe Trautteur and Raniero Virgilio set out to test this explanation. At the heart of their model lies the question whether Nelson's success is indeed best explained by his strategy or whether the superior firepower of his ships alone would have led to his victory. Trautteur and Virgilio built a model that considered, for each ship, local wind conditions, speed, firepower (with differing damage done by close, mid-range and long distance shots) and three basic statuses: attack, retreat and sink. The authors could rely on rather accurate historical data on wind conditions, the nature of the ships of both fleets, and the exact strategies applied by the two commanders. Overall, there are not too many factors involved and it is plausible that any external influences can be ruled out. Since this was a military operation, it can also be assumed that all ships followed the orders they were given and did not develop deviating strategies.

By reproducing the historical setting, they show that indeed Nelson's fleet would have been victorious with low casualties on the British side. They disagree with Lanchester's finding that without Nelson's peculiar strategy the battle had been lost. Instead they argue:

> ...our counterfactual simulations showed that English victory always occurs unless the environmental variables (wind speed and direction) and the global strategies of the opposed factions are radically changed, which lead us to consider the British fleet victory substantially ineluctable [33].

This means that Nelson's victory can be explained by his ships' superior firepower alone. His strategy, however, led to the extremely low casualty rate among the British. It caused the

> the Franco-Spanish vanguard to be substantially out of the fight for a long time, so as to minimize the risk for the English of losing ships and men [34].

Provided that the parameters of this model can indeed be trusted to adequately represent the relevant factors in historical naval warfare, it is a convincing case against Lanchester's theory. Moreover, the hypothesis that the superior firepower of the British fleet gave it a strong advantage to achieve victory in the first place means that Nelson's unconventional strategy was in fact only contributing to the relatively small number of casualties and not the outcome of the battle itself. Overall, the model falsified Lanchester's hypothesis, helped differentiate the evaluation of Nelson's strategy and, with the hint at the differing firepower, provided evidence for a new explanation.

6.4.3 The Flu in Manitoba

The fatal consequences of the Influenza epidemic in 1918/19 are well documented. Millions of people worldwide died as a direct consequence of the infection. This final case study offers a micro-historical analysis of the consequences of the epidemic in a remote area in Manitoba. The model reconstructs interactions between three neighbouring fur trade communities. One, Norway House, was affected by the flu and 18% of its inhabitants died as a consequence. The other two posts, God's Lake and Oxford House, however, did not report any deaths even though travels between all three were frequent [25, p. 759f]. How can this be explained?

The model discussed here evolved from two earlier attempts to model the fur trading communities in the area. Each of the trading posts was surrounded by a number of camps, essentially the hunting grounds of the hunters were they spent the winter together with their families. There is historical evidence that the flu was brought to the region by a single infected mailman. Reports by the Hudson Bay Company allow the reconstruction of travels to each of the three posts. Death rates and recovery times from the flu were adopted from other records in Canada and general medical observations [25, p. 759]. Number, age and gender of the agents could also be reconstructed from historical records.

Agents in the model "move randomly about their respective home locations and (...) travel to other locations according to a series of mobility rules and assumptions" [25, p. 760]. Analyses of the model produced differentiated results and some rejections of the findings produced by earlier versions of the model. In total however, the researchers draw the following conclusions:

> Travel between the posts during the winter was infrequent, and the journey took days. This, combined with the relatively short infectious period of influenza, meant that any infected traveller who managed to survive the journey from Norway House would probably not have been infectious upon arrival at Oxford House or at God's Lake [25, p. 764].

The authors state that they wish to further enhance their model. However, even at this stage they deliver a convincing answer for a question that could have hardly been answered without their computational model.

6.5 Discussion

Agent-based models provide an opportunity to formalize thinking about causalities of past processes and to critically re-assess underlying assumptions in a way previously limited to thought-experiments. In the words of Shawn Graham, "If this model clearly reflects my understanding of historical process X, then these are the natural consequences of that understanding" [12]. Given the rather small number of available models of historical events and the even smaller number of models that can present plausible results, it is difficult to compile a finite list of essential characteristics for models in the historical disciplines. Research in this field is without doubt still in its infancy and not yet well developed. Still, by comparing the approaches existing models took and by integrating experiences from models in other disciplines, a rough outline of what contemporary historical models share is possible:

(1) low number of influencing causal factors
(2) well-defined pre-existing hypotheses that can be tested
(3) stable, reproducible settings for environmental and human agents
(4) detailed information of the historical case study (the more the better)

Modelling military operations seems to be particularly promising: Examples like the Battle of Trafalgar and the Byzantine army on the march (not discussed here, see reference 11) have overall yielded good results beyond the falsification of competing explanations.

Most modellers of the simulations I investigate, and others like Joshua Epstein, emphasize that the most important insights gathered through modelling often occur during the process of building them. The forced rigidity of computer programs together with the need to capture the essence of interactions between agents and the need to simplify the model as far as possible prove to be highly productive.

This might be felt even more strongly by scholars of the historical disciplines who are used to concentrating on the specifics of an event and not on regular patterns. Arguably here lies the biggest challenge for any model: Just how much can you reduce historical events to rule-based behaviour? To which extent are unpredictable "odd ones out" essential for satisfying explanations? All of the case studies I analyse chose to design stable environments in which the rules of behaviour changed in pre-defined ways and struggled hard to match them to the observed historical outcome.

In his article "Why model?" Epstein presents 16 reasons to model. While all of them might apply in one way or the other, two stand out for their particularly close relationship to historical analysis[a]:

(11) Challenge the robustness of prevailing theory through perturbations

(12) Expose prevailing wisdom as incompatible with available data [10].

All case studies made use of models in this sense: The Anasazi research team successfully challenged the explanation that the communities left due to climate changes alone, known facts from the Battle of Trafalgar were used to differentiate earlier evaluations of Admiral Nelson's strategy, and the fur trading communities that were spared from the influence remained widely unaffected not due to sheer luck, but because of their remoteness and increased travel times during the winter. All of them tackled a pre-existing explanation for a historical event and used models based on available historical data to challenge it; some went on to present an alternative explanation. Arguably these can never be full explanations in a historical sense. They can, however, help to further an understanding of the subject of research and therefore contribute to explanations [37, p. 116].

Despite all the efforts to utilize historical data, all of the aforementioned models had to rely on plausible assumptions in order to calibrate their models. While these are not necessarily better than other implicit interpretations of facts, they have the advantage that they are, at least to the designer of the model, plainly

[a]Weber makes a similar point by stating that models in the historical disciplines will probably not be useful to find new explanations but in order to question existing ones and to the exclusion of unrealistic ones [34, p.116].

distinguishable from settings based on historical facts. Not only are they easily modifiable [similarly: 10; 11], in the process of analyzing historical narratives they also help us to distinguish better between historical facts, their interpretation and retrospect assumptions we invariably have to make.

Many historical agent-based models investigate dimensions of human exchange outside the socio-cultural-political sphere. This is most likely due to the inconsistency of human behaviour as opposed to infections, gunfire or changes in climate. In a review of the volume "Simulating Change: Archaeology into the Twenty-First Century" [3], Archaeologist Jim Doran specifies this problem which contributed to the so far "disappointing" results of modelling in the discipline:

> As regards the future, there is a deep further difficulty that is all too often overlooked. Distinctive human social structures and social processes emerge from distinctive human cognition. But we do not yet know how to model human cognition on a computer in other than relatively superficial and oversimplified ways. Thus we cannot yet experiment with the models that really matter: those that capture more than simple routine cognitive behavior. Archaeology faces this challenge as do all the social sciences. For help we need to look to developments in artificial intelligence engineering and in cognitive science modeling [6].

Modelling history has been and will remain a challenge due to the scarcity of available data and the nature of the discipline and its research interests. Case studies including the three that were exemplified here, however, do show that progress is being made and that simulations can under certain conditions yield more convincing results than traditional source analysis alone.

References

1. Anreiter, Peter, ed. *Mining in European history: Special conference of the SFB HiMAT: Historical Mining Activities in the Tyrol and Adjacent Areas: Impact on Environment and Human Societies: Conference Guide, Innsbruck, November 12–November 15, 2009*. Innsbruck: University Press, 2009.
2. Carpenter, Connie, and Lisa Sattenspiel. "The design and use of an agent-based model to simulate the 1918 influenza epidemic at Norway House, Manitoba." *Am J Human Biol*, **21**(3), 2009: 290–300.

3. Costopoulos, Andre, and Mark Lake. *Simulating Change: Archaeology into the Twenty-first Century*. Salt Lake City: University of Utah Press, 2010.

4. Dean, Jeffrey, et al. "Understanding Anasazi culture change through agent-based modeling". In *Dynamics of Human and Primate Societies: Agent-Based Modeling of Social and Spatial Processes* (ed. Timothy A. Kohler and George J. Gumerman). New York, Oxford: Oxford University Press, 1999.

5. Diamond, Jared. "Life with the artificial Anasazi". *Nature* **3**(419), 2002: 567–569.

6. Doran, Jim. "Iruba: an agent-based model of the guerrilla war process". In *Representing Social Reality: Pre-proceedings of the Third Conference of the European Social Simulation Association (ESSA), Koblenz, Sept 5–9, 2005* (ed. Klaus Troitzsch), 198–205. 2005. http://cswww.sx.ac.uk/staff/doran/34.PDF.

7. Doran, Jim. "Review of simulating change: archaeology into the twenty-first century (Foundations of Archaeological Inquiry)". *J Artif Soc Soc Simulation,* 14(4), 2011. Accessed February 5, 2012. http://jasss.soc.surrey.ac.uk/14/4/reviews/2.html.

8. Düring, Marten, "How reliable are centrality and clustering measures for data collected from fragmentary and heterogeneous historical sources? A case study". *Proceedings of Conference "The Connected Past. People, Networks and Complexity in Archaeology and History"*, March 2012 in Southampton, 2014.

9. "Personal website". Accessed October 17 2012, http://martenduering.com.

10. Epstein, Joshua M. "Why model"? *J Artif Soc Soc Simulation,* **11**(4), 2008. Accessed February 5, 2012.

11. Graham, Shawn. "Agent-based modeling, archaelogy and social organisation: the robustness of rome". *Archaelogical Computing Newsletter,* **63**, 2005: 1–6.

12. Graham, Shawn. "Behaviour space: simulating roman social life and civil violence". *Digital Studies/Le champ numérique,* **1**(2), 2009. Accessed February 5, 2012. http://www.digitalstudies.org/ojs/index.php/digital_studies/article/view/172/214.

13. Ierodiakonou, Katerina, and Sophie Roux. *Thought Experiments in Methodological and Historical Contexts*. Leiden; Boston: Brill, 2011.

14. Janssen, Marco. "Understanding Artificial Anasazi". *J Artif Soc Soc Simulation*, **12**(4), 2009. Accessed February 5, 2012. http://jasss.soc.surrey.ac.uk/12/4/13.html.

15. Kee, Kevin, Shawn Graham, Pat Dunae, John Lutz, Andrew Large, Michel Blondeau, and Mike Clare. "Towards a theory of good history through gaming". *Can Hist Rev*, **90**(2), 2009: 303–326. doi:10.3138/chr.90.2.303.

16. Kohler, Timothy A., and George J. Gumerman, eds. *Dynamics of Human and Primate Societies: Agent-Based Modeling of Social and Spatial Processes*. New York, Oxford: Oxford University Press, 1999.

17. Kohler, Timothy A., George J. Gumerman, and Robert G. Reynolds. "Simulating ancient societies". *Scientific American*, July 2005. http://www.scientificamerican.com/article.cfm?id=simulating-ancient-societ.

18. Kohler, Timothy A., and Mark D. Varien. *Emergence and Collapse of Early Villages: Models of Central Mesa Verde Archaeology*. Berkeley: University of California Press, 2012.

19. Kowarik, Kerstin, Hans Reschreiter, and Gabriel Wurzer. "Modeling the Bronze Age salt mines of Hallstatt". In Mining in European History: Special Conference of the SFB HiMAT: *Historical Mining Activities in the Tyrol and Adjacent Areas: Impact on Environment and Human Societies: Conference Gui*de, Innsbruck, Nov. 12–Nov. 15, 2009. (ed Peter Anreiter), 44–53. Innsbruck: University Press, 2009.

20. Lorenz, Chris. Konstruktion der Vergangenheit: *Eine Einführung in die Geschichtstheorie*. Köln: Böhlau, 1997.

21. Malerba, Franco, Richard Nelson, Luigi Orsenigo, and Sidney Winter. "History-friendly models: an overview of the case of the computer industry". *J Artif Soc Soc Simulation*, **4**(3), 2001. Accessed February 5, 2012. http://jasss.soc.surrey.ac.uk/4/3/6.html.

22. Mey, Tim de, and Erik Weber. "Explanation and thought experiments in history". *Hist Theory*, **42**(1), 2003: 28–38.

23. Murgatroyd, Philip, Bart Craenen, Georgios Theodoropoulos, Vincent Gaffney, and John Haldon. "Modeling medieval military logistics: an agent-based simulation of a Byzantine army on the march". *Comput Math Org Theory*, 1–19.

24. Newman, Christopher. "The Kentucky Dark Patch Knight Riders' rebellion". *New Eng J Polit Sci*, **1**(2), 2006: 3–28. http://www.northeastern.edu/nepsa/journal/archives/the_new_england/.

25. O'neil, Caroline A., and Lisa Sattenspiel. "Agent-based modeling of the spread of the 1918–1919 flu in three Canadian fur trading communities". *Am J Human Biol*, **22**(6), 2010: 757–767. doi:10.1002/ajhb.21077.
26. *Play the Past*, available at www.playthepast.com (10.1.2012).
27. Preiser-Kapeller, Johannes. "Working paper 'Historical Dynamics of Byzantium'". Unpublished manuscript, http://www.oeaw.ac.at/byzanz/repository/Preiser_WorkingPapers_Calculating_I.pdf.
28. Ritterfeld, Ute, Michael Cody, and Peter Vorderer. *Serious Games: Mechanisms and Effects*. New York (N.Y.); London: Routledge, 2009.
29. Sattenspiel, Lisa, and Alun Lloyd. *The Geographic Spread of Infectious Diseases: Models and Applications*. Princeton: Princeton University Press, 2009.
30. SimulPast. "SimulPast. Simulating the past to understand human behaviour". Accessed February 5, 2012. http://www.simulpast.es/.
31. Tetlock, Philip E., Richard N. Lebow, and Geoffrey Parker, eds. *Unmaking the West: "What-If" Scenarios That Rewrite World History*. Ann Arbor: University of Michigan Press, 2006.
32. Tetlock, Philip E., and Geoffrey Parker. "Counterfactual thought experiments: why we can't live without them and how we must learn to live with them". In *Unmaking the West: "What-If" Scenarios That Rewrite World History* (Ed. Philip E. Tetlock, Richard N. Lebow and Geoffrey Parker), 14–46. Ann Arbor: University of Michigan Press, 2006.
33. Troitzsch, Klaus, ed. *Representing Social Reality: Pre-proceedings of the Third Conference of the European Social Simulation Association (ESSA), Koblenz, Sept 5–9, 2005*. 2005.
34. Trautteur, G., and R. Virgilio. "An agent-based computational model for the battle of Trafalgar: a comparison between analytical and simulative methods of research". In *WET ICE 2003: Proceedings*, 377–382. [Piscataway, NJ]: IEEE, 2003. http://www.computer.org/portal/web/csdl/abs/proceedings/wetice/2003/1963/00/19630377abs.htm.
35. Turchin, Peter. *Historical Dynamics: Why States Rise and Fall*. Princeton, NJ [u.a.]: Princeton University Press, 2003.
36. van den Hoven, M., A. van den Bosch, and K. Zervanou. "Beyond reported history: strikes that never happened" In *Proceedings of the First International AMICUS Workshop on Automated Motif Discovery in Cultural Heritage and Scientific Communication Texts* (ed. S. Daranyi and P. Lendvai), 20–28. Vienna 2010.

37. Weber, Karsten. *Simulation und Erklärung: Kognitionswissenschaft und KI-Forschung in wissenschaftstheoretischer Perspektive*. Münster: Waxmann, 1999.

38. Weber, Karsten. "Erklärung historischer Abläufe mit Computersimulationen". *Hist Soc Res/Hist Sozialforschung*, **32**(4), 2007: 94–121.

PART 2

COMPLEX ADAPTIVE SYSTEMS IN THE SOCIAL SCIENCES

Chapter 7

Complexity Theory and Political Change: Talcott Parsons Occupies Wall Street

Martin Zwick

Systems Science Graduate Program, Portland State University, 1604 SW 10th Ave, Portland, OR 97207-0751, USA

zwick@pdx.edu

Complexity theory can assist our understanding of social systems and social phenomena. This chapter illustrates this assertion by linking Talcott Parsons' model of societal structure to the Occupy Wall Street movement. Parsons' model is used to organize ideas about the underlying causes of the 2008 U.S. recession. While being too abstract to depict the immediate factors that precipitated this crisis, the model is used to articulate the argument that vulnerability to such events results from flaws in societal structure. This implies that such crises can be avoided only if, in Parsons' terms, structural change occurs in the relations between polity, economy, community, and culture. The Occupy movement has called attention to the need for such fundamental change.

7.1 Introduction

This chapter revisits an early cybernetic and systems-theoretic model—today it might be called a complex systems model—

Complexity and the Human Experience: Modeling Complexity in the Humanities and Social Sciences
Edited by Paul A. Youngman and Mirsad Hadzikadic

ISBN 978-981-4463-26-3 (Hardcover), 978-981-4463-27-0 (eBook)
www.panstanford.com

proposed by the sociologist Talcott Parsons [15–16] and argues that this model can help us understand some of the underlying causes of the 2008 recession that continues, as of 2013, to afflict the US economy. The recession, which followed a near total meltdown of the US financial system, has involved massive losses of jobs, homes, and savings. This chapter does not focus on concrete and proximal causes of this crisis, i.e., on actions by home owners, mortgage providers, banks, investments firms, regulatory agencies, Congress, the President, etc., about which there is no shortage of accounts. Instead, it poses the question of why the United States was vulnerable to such a crisis, more specifically, what aspects of the US societal structure generated this vulnerability.

By approaching the recession crisis abstractly, Parsons' model also sheds light on past crises, whose concrete and proximal causes were different, as well as illuminating societal problems of a completely different character. Finally, the model can be used not only to diagnose problems, but also to offer solutions. Specifically, it suggests that changes in societal structure might reduce vulnerability to such economic disasters.

To be succinct, the model says: (i) A modern society, as distinct from a pre-modern traditional society, is *differentiated* in that economy, polity, community, and culture are distinguishable from, but in interaction with, one another. (ii) In general, differentiation of a whole can be flawed: different aspects of a system can be linked together too weakly or too strongly, or links can be of the wrong kind, or one component can unduly influence other components. (iii) The 2008 US political-economic crisis and similar crises in the past are arguably the result of flawed differentiation, i.e., relations between economy, polity, community, and culture are at least partially dysfunctional. This may be the underlying cause of many (but not all) of the crises endemic to Western societies. These crises are systemic, and avoiding them requires structural change.

Although the discussion that follows has a politically liberal orientation, Parsons' model itself is neutral, that is, using the same model, another theorist with a different political orientation might arrive at different conclusions. Indeed, this model was once characterized as being inherently conservative, assuming stability and functionality for societal structures. This characterization, however, was incorrect, since Parsons' structural functionalism can actually be used to explain either stability or instability and either functionality or conflict.

Section 7.2 describes Parsons' model. Section 7.3 discusses the flaws—as depicted by this model—that characterize the societal structure of the United States. Section 7.4 considers some proposals for political change from the *99 Percent Declaration* of the Occupy Wall Street [14] movement and that, if implemented, might rectify some of these flaws. This *Declaration* is not an official set of demands—the Occupy movement has been reluctant to put forward such demands—but rather the output of a working group. Regardless of its lack of official status, the document is of interest here because its demands address some of the structural flaws that exist in the US societal system. The structural changes called for in these demands are not original to the OWS movement but neither have they been prominent in political debate. Indeed, their recent modest gain in prominence is at least partially due to the Occupy movement.

Both the flaws of differentiation whose existence is asserted in Section 7.3 and the proposed corrective measures discussed in Section 7.4 can be—and have been—discussed without any reference to Parsons' model. The point of this chapter is to suggest that this model offers a coherent framework for talking both about problems connected with the recession and possible solutions. By expressing them in the language of Parsons' model, these problems are seen as the "lawful" result of an underlying structure, as opposed to merely historical contingency. Similarly, from the perspective of this model, many proposed solutions to these problems derive from the need for structural change, and are not merely ad hoc remedies. Indeed, several demands in the OWS *Declaration* aim precisely at those structural changes called for by the Parsonian analysis.

7.2 Parsons' Model of Societal Systems

Talcott Parsons' [15–16] "action system" is a theoretical framework for modeling societal phenomena. While the model is an old one, it would be a mistake to think that early work in systems theory/complexity is necessarily less relevant to contemporary social issues than more recent work. As noted above, the model does not assume harmony between the differentiated components of societal systems; this paper, in fact, uses it to focus on disharmonies. These disharmonies can cause change, or, as in the current US societal structure, they can be locked in if the system is stuck in a suboptimal state. The analysis of structure not only allows one to explain function and account for stability; it also allows one to explain dysfunction and account for change, or the absence of change. As an explanation of dysfunction—such as an economic recession—it also has normative implications.

"Action," as a general concept, is said by Parsons to have the tetradic structure shown in Fig. 7.1a. With minor modifications of Parsons' ideas, this structure is a special case of Bennett's [6] tetradic system of purposeful action shown in Fig. 7.1b. Parsons and Bennett use the word "goal" in different ways: Parsons uses the word, which labels the "east" component of Fig. 7.1a, to mean a specific (tactical) purpose; Bennett uses the word, which labels the "north" component of Fig. 7.1b, to mean a more general (strategic) objective.

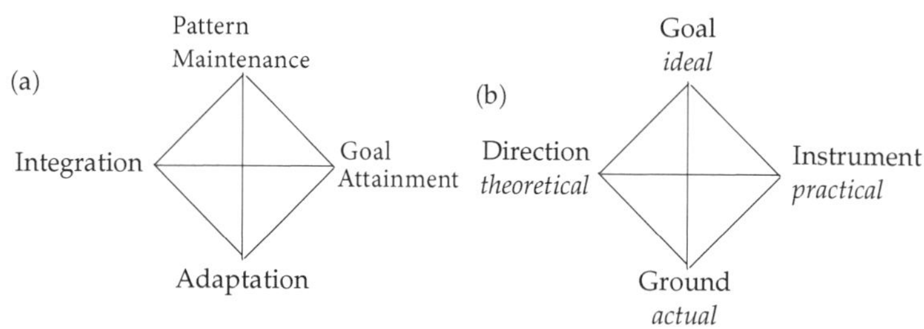

Figure 7.1 (a) Parsons' action model. (b) Bennett's similar model of purposeful action.

In Parsons' model, action aims at the general goal of pattern maintenance. It does so using basic processes of adaptation, the ground of action, which relate the system to its environment. These processes are used to attain specific goals by an instrument of action, which is given direction by a component promoting integration. This tetradic scheme is known as Parsons' AGIL model: A for adaptation, G for specific goal attainment, I for integration, and L for the maintenance of latent patterns.

Applied to societal systems, this scheme yields Fig. 7.2. The economy is the ground of societal action; it provides the material processes needed to adapt to internal and external conditions. The polity, or political order, is society's instrument for attaining specific goals. The community[a]—relations based on kinship, ethnicity,

[a]Parsons' actual phrase, "societal community," is shortened here to "community."

language, location, interests, etc., often analyzed with a focus on race, class, and gender—is the integrating component of the societal system and provides the direction for the polity as instrument. Culture, or the conglomeration of religion, science, the media, literature and arts, constitutes the patterns of knowledge and values that organize and are maintained by societal action. Figure 7.2 includes the environment whose link to the economy is not shown.

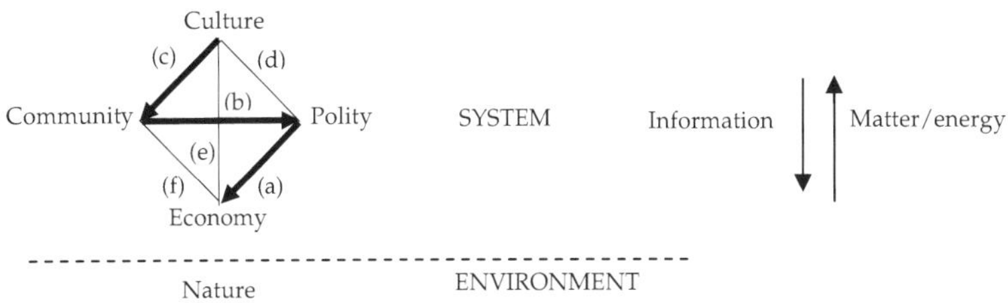

Figure 7.2 Parsons' action tetrad, applied to societal systems.

Economy, polity, community, and culture are the subjects, respectively, of the social science disciplines of economics, political science, sociology, and anthropology; for Boulding [8], they constitute the "exchange system" (economy), the "threat system" (polity), and the "integrative system" (community, culture). The simplicity and generativity of Parsons' model, its correlation with conventional ways of parsing societal systems, and its relation to a more general model like Bennett's tetrad all recommend this model as a framework for discussing societal systems and their pathologies.

Using ideas from cybernetics and systems theory, Parsons orders the four components of the system hierarchically, indicated by the zigzag sequence of directed links shown in bold in the figure. Culture, the top component, is information-like; economy, the bottom component, is matter-energy-like.[b] Informational regulation occurs downward following the zigzag sequence of links: (a) polity regulates economy, (b) community directs polity, and (c) culture guides community. There is an upward flow as well; the lower levels provide matter-energy support for the higher ones. In Marxist terms, the economy is the base while the other components are the superstructure. The Marxist emphasis on determination from below

[b]The economy has informational as well as matter-energy aspects, but relative to the other components its matter-energy character is salient.

[13] is incomplete: there is determination both from above and from below. In fact, the upward flow of energy has its ultimate source not in the economy, but in nature. The flux of matter-energy through a societal system organizes it [1], the flux providing the negentropic basis for its physical infrastructure. In this dual flow, this scheme accommodates both materialist and idealist views: materialist views are reflected in the upward flows; idealist views, in the downward flows. This is still only a partial representation of interactions in the structure. There are upward information flows as well, e.g., community generates culture, as well as other non-hierarchical interactions.

Intra-component interactions (not shown in the diagram) are stronger in the lower components of economy and polity than in the higher components of community and culture. One can thus simplify the tetrad into the dyad of economy-polity versus community-culture as was done by Habermas [10], who speaks of the "system"—economy plus polity—as opposed to the "lifeworld"—community plus culture. Habermas argues that because of the strength of economic and political forces, the outwardly oriented system "colonizes" the inwardly oriented lifeworld. However, this aggregation oversimplifies the tensions that can exist within the tetrad, ignoring the important tension between polity and economy that is central to the subject under discussion.

The upward and downward flows do not exhaust the connections between the components. The four components are interconnected in other ways, and the set of dyadic links are labeled (a) through (f) in Fig. 7.2. The links have multiple meanings. To illustrate, "modernity" in Western societies means that:

(a) the economy is partially autonomous and partially controlled by the polity;
(b) civil society determines and participates in the political order;
(c) cultural values guide but do not legally constrain private activity;
(d) church is separated from state (though religious values influence community);
(e) culture is independent of, yet supported by, commercial life;
(f) interpersonal relations are not dominated by those of economic exchange.

The idea that "modernization is differentiation" means that differentiation increases in modernity, not that differentiation is total. In the terminology of Herbert Simon [18], a societal system is only "partially decomposable," where being "decomposable"[c] means being reducible to a sum of parts. Decomposability is partial since components constrained by links with other components are not fully autonomous. In modern societal systems, differentiation is balanced by integration. The economy and the polity need to be connected, since economies are not adequately self-regulating and do not necessarily produce results in accord with societal values. The polity and community need to be linked according to principles of democracy and civic responsibility. The community needs to be guided by coherent values. Other types of links are also important.

The simplest links, as exemplified above, are dyadic, but links can involve more than two components [20]. For example, there might be a triadic interaction between polity, community, and culture that cannot be decomposed into dyadic relations. There can be interactions that are not separable at all into simpler ones. For a system with only four components, a four-way interaction that cannot be fully decomposed can be called "holistic." Archaic societies were more holistic than modern societies in that economy, polity, community, and culture were more tightly integrated and less distinguishable. With modernity, these components became more autonomous and distinct, but only partially so. It is impossible to conceive of a society where these components are completely independent, although some still harbor fantasies of a completely autonomous market.

Structural analysis goes beyond considering possible links between components. Each component has an inner structure consisting of sub-components of the same four types; this structural similarity at different scales is referred to as being "fractal." Thus, in the AGIL system, A can be further decomposed into A, G, I, and L *sub*-components; similarly for G, I, and L. For example, community, which is the I component of the societal system, has a polity-like (G) sub-component; this G within I, is "citizenship," with its political opportunities and obligations.

---

[c]"Decomposability" is a formal systems-theoretic notion that encompasses both differentiation and disintegration, where "differentiation" has a biological or social system meaning, and does not refer to mathematical expressions like dx/dt.

This fractal conception is theoretically elegant, but a more natural decomposition of the US polity, for example, is the separation of powers among the executive, legislative, and judicial sub-components. This triadic differentiation of the polity was an explicit intention of the founding fathers, and the institutionalization of separation of powers is an example of a specific and conscious—and partially successful—solution to a problem of differentiation. The polity is also decomposed hierarchically and spatially, and the framers of the constitution also grappled explicitly with this different challenge of differentiation, namely the issue of federal versus states rights and powers.

Similarly, while the economy might be decomposed into AGIL sub-components, a more natural decomposition might be based on the distinction between information and the matter-energy aspects of the economy. The financial and knowledge sectors of the economy are informational, while production and distribution of material goods and energy is obviously matter-energy-like. Service industries may resemble either of these or be intermediate in character.

7.3 Flaws of Differentiation

As already noted, the functionalism of Parsons' model does not imply a natural harmony between its components, which can be mismatched or improperly related. This is arguably the case for US society. The model posits a downward flow of information, i.e., regulatory control, from culture to community to polity to economy, and the most critical segment of this pathway is the last link: the control of the economy by the polity. In the United States today, the direction of control in the last link is substantially in the opposite direction: the economy, i.e., the large corporations and powerful special interests, largely controls the polity. This is accomplished by lobbying, by benefits provided to officials, by campaign contributions, and by the revolving door through which individuals move from public service to private employment that capitalizes on contacts and knowledge gained while in public service. The failure of the polity to prevent the recession and deal with it effectively after its onset is partially due to this reversal of direction of influence.

Faulty differentiation is not merely a matter of direction of control. Loss of control of the economy by the polity allowed behavior in the financial sector that precipitated the recession and the danger of system-wide collapse that was barely averted. Regulation of the financial sector having been dismantled, banks and investment firms indulged in risky behavior. While the financial crisis led to temporary restoration of government influence on these firms and to the recognition of the systemic dangers of unregulated financial activity, it is doubtful that the proper lessons from this crisis have been learned, as adequate oversight of this sector is still not instituted.

As mentioned above, one way that the economy exercises its control over the polity is via campaign contributions. The Supreme Court in the Citizens United case decided that corporate contributions are protected free speech. This stems in part from the legal status as "persons" that corporations now enjoy in their use of money to influence the political process. From the perspective of Parsons' model, this legal status is anomalous and exemplifies another structural flaw: personhood naturally applies to individuals in the community, not to private organizations in the economy. The right of free speech is critical to the community–polity link, which is the basis of democracy. Conferring upon corporations a right to "free speech" usurps and dilutes a right that legitimately belongs to a different component of the structure. "One person, one vote" reflects proper control of the polity by the community, as represented by the community → polity link in Fig. 7.2. "One dollar, one vote," however, reflects a distortion of this principle, one that results in dominance of the polity by the economy, reversing the directionality of the polity → economy link in Fig. 7.2, which is normative for Parsons' theory.

In any differentiated structure involving active agents, different components vie for influence on the whole. Where there are multiple components, there is rarely a single organizing principle that governs the system, but usually two or more principles that conflict with one another. Contradiction—or tension—is thus built inherently into the system. US society is often described as an example of "democratic capitalism." This does not have to be an oxymoron: democracy refers to the community–polity link, and capitalism refers to the economy component and the polity–economy link. These two organizing principles might be harmoniously related or they might be at odds, but if they are at odds, as is often the case, one must have priority. In Parsons' model, informational regulation flows downward from community to polity to economy, so in principle, democracy, not capitalism, has priority. However, the actuality in the current US system is arguably the reverse of this.

In von Bertalanffy's [7] terminology, a "leading part" of a network—in contemporary parlance its main "hub"—is a component that has a dominant influence on other components. Because causality in a network is invariably mutual, i.e., circular, one cannot make simple distinctions between causes and effects, so the notion of a leading part plays a role similar to that of a primary cause. In the US system, the economy is the leading part, as shown in Fig. 7.3a. It dominates not only the polity, but also community, culture, and nature. This resembles the Marxist view of the economy as a base that determines the rest of society as a superstructure, but from a systems—and Parsonian—perspective, this structure is not preordained for the differentiation that comes with modernity. It is a distortion, and one that can be corrected even without replacing the economy by the polity as a leading part.

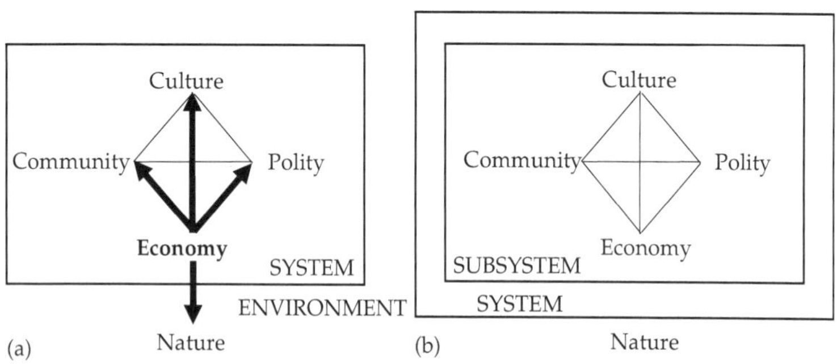

Figure 7.3 Economy as leading part—or subsystem of a subsystem. Arrows indicate directions of control. (a) The current US system. (b) The "true" relation of any economy and societal system to nature.

Consider the economy–community link. While economic theory recognizes that economies function well only when negative externalities are internalized by decision-making agents, few mechanisms exist in a weakened polity that force externalities to be internalized. In the 2008 recession, irresponsible, but in most cases legal, acts by many corporations resulted in enormous harm to the community, yet the community cannot extract compensation for this harm from those who caused it. Moreover, the government bailed out some of the offending firms but gave the community only minimal help via stimulus programs and unemployment benefits. There are also longer-term externalities affecting the community that manifest in the large and growing income inequality between the great majority and a privileged minority: to use the simplifying and expressive terminology of the Occupy movement, between the 99% and the 1%.

So far the analysis has focused on links between the four components of Parsons' model. Tensions and distortions can exist also within these components. For example, the differentiation of the polity into executive, legislative, and judicial branches can produce deadlock when executive and legislative branches are controlled by different parties. In the Obama administration, this indeed blocks effective action by the polity, but this is not the core of the crisis, since it is doubtful that even a Congress controlled by a filibuster-proof Democratic majority would implement structural changes needed to prevent similar crises in the future.

Intra-component distortions in the economy are also relevant to recession events. The economy is differentiated into informational versus matter-energy sectors. The informational sector—especially its financial part—has grown anomalously large relative to the matter-energy sector. Although economic theory ascribes to the financial sector the important function of resource allocation, it is doubtful that the added value that this sector contributes by performing this function justifies the amount of wealth it siphons off for itself. Here the situation is the reverse of a matter-energy-like "system" colonizing an information-like "lifeworld" (to use Habermas' terms): In the economy, it is part of the informational sector that is parasitic on the rest of the economy and on the community.

Such parasitism notwithstanding, the government bailed out offending firms judged "too big to fail." That is, not only is the financial sector anomalously large, but some firms within it are also too big. The restoration of some regulatory powers and the assumption of new powers to prevent system-wide collapse are unlikely to remedy the dangers that inhere in large size. The corrective measures taken so far will not insulate regulatory agencies from capture or prevent recurrence of crisis triggering bubbles. Already, firms too big to fail have become still bigger.

The economy may also be too integrated, so disturbances in parts of it can spread and become amplified. It also generates bubbles, which shows that it is not optimally self-regulating. The real estate bubble was an important cause of the recession, and before this bubble there were many others. The tendency to generate bubbles is a structural flaw in that the positive feedback that manifests in speculative behavior is not adequately controlled by the negative feedback that characterizes ideal markets. From one viewpoint cyclic behavior is just an attractor that many dynamic systems exhibit, but from another point of view [5], cyclicity demonstrates the existence of a "logical contradiction" within the system.

7.4 Digression

Structural flaws are connected also with other, even more critical societal problems. The dominance of the economy over the polity limits our ability to protect the biosphere against harm from industrial activity—from climate change to depletion of non-renewable resources to species extinction, etc. If allowed to do so, the economy takes control of nature, which it treats as mere "environment" as shown in Fig. 7.3a, i.e., as a source for resources and a sink for wastes, both incorrectly assumed to be infinite. The relation that economies actually have with nature is depicted in Fig. 7.3b: An economy is really a subsystem nested within a subsystem that completely depends on the biosphere. Societies are wholly owned subsidiaries of nature; even if we insist on the illusion that the reverse is true, nature "belongs" to the community, not to the economy or polity. Protection of nature may require new community-nature links, in which natural commons like air, water, etc. are held in trust for future generations as part of a public sphere that is partially autonomous from both economy and polity. Community is not only the locus of societal integration; it is the component via which the future of the society can compellingly present its needs to the present.

On a much smaller scale, a second example of a structural flaw that indirectly relates to the recession and aggravates its consequences is one aspect of the community–economy link, namely the connection that now exists between employment and health insurance. This link is a historical artifact that today is irrational, just as it would be irrational if children's access to education depended on their parents' employment. While the optimal roles that the community, polity, and economy should play in the provision of health care may be debatable, it is plain that health care and education do not entirely belong to the domain of the market.

This example and the previous one highlight the fact that Parsons' model is useful not only for talking about economic recessions but also about other issues confronting society today. One might add, in passing, that the 2012 Euro crisis obviously derives from the failure of the European Union to institutionalize necessary community–polity and polity–economy links.

7.5 Fixing the Flaws

As mentioned in the Introduction, the *99 Percent Declaration* of the Occupy movement contains demands that address some of the structural flaws discussed in the preceding section. While this document is not an official declaration of the movement, it succinctly and directly raises fundamental issues for public discussion. From the joint perspective of Parsons' model and OWS demands, if economic crises are to be avoided in the future, the central challenge is to establish the right direction of control between polity and economy. This requires both increasing and decreasing the separation between these two components. With respect to economy → polity influences, separation needs to be increased; with respect to polity → economy influences, separation needs to be decreased. These changes, plus others discussed below, are summarized in Table 7.1.

Table 7.1 Needed structural changes

Parsons' model	Some occupy demands (and other correctives)
Economy → polity	Insulate guardian functions from commercial influence
Polity → economy	Empower regulation keyed to scale
Economy → community	Decrease and internalize negative externalities of the economy
Community → polity	Election reform; regulatory transparency

With respect to economy→polity interactions, a sharp distinction needs to be made between the "guardian" values of the polity and the "commercial" values of the economy [12]. Plato [17] long ago observed that mixing these deeply different societal functions invariably leads to corruption, and advocated radical measures to prevent this. This separation is forcefully called for in the *Declaration* (without using the guardian-commercial terminology):

(3) *Elimination of All Private Benefits and "Perks" to Politicians. The 99% of the American People demand the immediate prohibition of special benefits to all federal public employees, officers, officials or their immediate family members. Public officials, politicians and their immediate families shall be banned from ever being employed by any corporation, individual or business that the public official specifically regulated while in office. No public employee, officer, official or their immediate family members shall own or hold any stock or shares in any corporation or other entity that the public official specifically regulated while in office until a full 5 years after their term is completed.*

There shall be a complete lifetime ban on the acceptance of all gifts, services, money or thing of value, directly or indirectly, by any elected or appointed public official or their immediate family members, from any person, corporation, union or other entity that the public official was charged to specifically regulate while in office. The term 'specifically regulate' shall mean service on a committee or sub-committee or service within any agency or department of the federal government responsible for the regulation of the person, union, corporation or entity seeking to directly or indirectly confer a benefit to a public official.

It may well be that only radical measures of this sort, which enact Plato's views into law, can achieve the separation of economy and polity that is necessary to avoid corruption. Such measures may require a constitutional amendment. This may still be insufficient to alter the power balance between polity and economy. Public funding of elections may thus also be needed, since the prohibition of campaign contributions from corporations and unions would not preclude contributions from wealthy individuals and well-funded political groups, which are more legitimate, being part of the community–polity rather than the economy–polity link. Another demand of the *Declaration* calls for this measure to reduce the political influence of big money:

(1) *Elimination of the Corporate State ... Private funding of campaigns from concentrated sources of wealth has corrupted our political system. Therefore, all private funding of political campaigns shall be replaced by the fair, equal and total public financing of all federal political campaigns ...*

It has been estimated that 94% of all federal political campaigns are won by the candidate who spends the most money. Our elected representatives spend far too much of their time fundraising for the next election rather than doing the People's business. This constant need for more and more money, causes our politicians to labor under conflicts of interest that make it impossible for them to act in the best interests of the American People.

The dominance of the polity by the economy also derives in part from the legal status of corporations as "persons," which as noted above usurps a status that properly belongs only to members of the community. This status needs to be revoked, and this also is a salient demand of the Declaration:

(2) *Abrogation of the "Citizens United" Case. The immediate abrogation, even if it requires a Constitutional Amendment, of the outrageous and anti-democratic Supreme Court holding in Citizens United v. Federal Election Commission. This heinous decision proclaimed by the United States Supreme Court in 2010 equates the payment of money to politicians by corporations, wealthy individuals and unions with the exercise of protected free speech. We, the 99% of the American People, demand that institutional bribery never again be deemed protected free speech.*

With respect to interactions represented in Fig. 7.2 by the polity→economy link, the changes needed are the opposite of the changes needed in the economy→polity link, namely the components need to be more tightly linked. The polity must have the powers necessary to regulate both the information, e.g., financial, and the matter-energy sectors of the economy, the former especially to prevent financial collapse of the sort that almost occurred in the

2008 crisis and to limit appropriation of wealth by the financial sector to what is justified by the function that it serves, the latter to address dangers of pollution, resource depletion, and species extinction. Regulation must be able to force economic entities to internalize negative externalities on the community and nature; by such internalization, market forces are mobilized toward reducing these externalities.

As to specifics of reregulation of the financial sector, the *Declaration* includes some proposals in items #16, "Banking and Securities Reform" and #18, "Ending the Fed." From a structural perspective, innovations in financial transactions ought to be assessed by regulators in terms of the function of resource allocation that theory credits this sector with performing. For example, it is doubtful that high speed computerized trading contributes positively to this function; unless such a positive contribution can be demonstrated, regulation should limit this activity because of the instability that it causes.

More generally, regulation needs to depend on scale, and this principle needs to be accepted as a central consideration that should govern the polity → economy link. The larger the impacts an economic entity can have, the more transparent its actions must be to public scrutiny, and the more tightly the entity must come under regulatory control. Very large and powerful economic entities that can massively impact community should be illegal; for such entities, only a principle of "too big to fail = too big to exist" can prevent similar crises in the future. A large size that poses systemic risk should be even more unacceptable than one that stifles competition. Alternatively, such large entities should simply be nationalized. Small economic units, however, should not be afflicted by onerous bureaucratic requirements. This requires vigilance since bureaucratic hypertrophy is a characteristic dysfunction of the polity. Regulation itself should be fully transparent to public scrutiny.

Even if the sizes of economic entities were limited, the economy is so highly interconnected that small disturbances can still have systemic effects. When an adverse system-wide event occurs, its causes are only superficially the specific disturbances that triggered it (e.g., the collapse of the real estate market), but more fundamentally the critical state of the system that amplified and spread the local dysfunction. Such an underlying vulnerability can be mitigated to some extent by limiting leverage and perhaps also by reducing direct dependencies between firms, but interconnectivity is a mark of mature economies, and indirect links may suffice to spread disturbances throughout the system. It would thus be desirable to reduce not only the disproportionate size of economic units but also their interconnectivity, but it is unclear how this could be accomplished within a regulatory mandate. It is also unclear if regulation could inhibit the tendency of the economy to produce bubbles; limiting their size to reduce negative externalities and to avoid the moral hazard of bailouts may be the best that can be done. Apart from the problem of giving regulatory agencies the legal powers needed to do their job, there is the problem of providing them with effective tools with which to exercise such powers, complicated by the fact that the theory used by regulators, policy makers, and economic agents is very likely to be inadequate to this job; witness the failure of the theory of financial risk widely accepted prior to the recession.

Weakening the economy→polity link and strengthening the polity → economy link may need to be supplemented by also strengthening the community → polity link. Two demands aimed at doing so are included in the *Declaration*: (1) term limits (item #4 in the *Declaration*), and (2) abolition of the Electoral College and other campaign finance and election reform (item #19). These issues are too complex to be taken up in this paper, but from the perspective of Parsons' model, the goal in community → polity changes should be to increase the accountability of the polity without diminishing its efficacy, and to clearly establish the primacy of the organizing principle of democracy over the organizing principle of capitalism.

There is also a need for change within culture. The prevalence of false beliefs about the self-regulating capabilities of markets, which is the consequence of the colonization of culture by the economy, needs to be counteracted. Public awareness that instability and dysfunction are as natural to economies as stability and optimality would enhance the legitimation of the polity → economy link. Economic regulation keyed to scale needs to be understood as the way to preserve, not destroy, markets. And while the distorted view of nature as resource—Fig. 7.3a as opposed to Fig. 7.3b—has been partially corrected by the growing recognition of the valuable "services" that ecosystems perform, a more radical understanding of the place of human societies in the biosphere is still needed to prevent collapse of a fossil-fuel-dependent bubble that is not merely economic but civilizational.

Finally, it needs to be noted that the Bennett tetrad also points to flaws in the Occupy approach: there is no logical connection between the signature tactics of OWS—the occupation of public spaces—and the demands of this movement, i.e., between instrument and goal, at least as expressed in the unofficial manifesto. There is also no well-articulated theory that connects tactics and objectives.

7.6 Summary

Theories of complexity—in older terminology, systems and cybernetics theories—can augment the intellectual resources of the disciplines of the social science and history. Parsons' model, a synthesis of sociological and early systems theory, offers what Gell-Mann [9] called "a crude look at the whole," here a societal whole. Jane Jacobs's revival of Plato's distinction between "guardian" and commercial ethics is systems-theoretic in character. Moreover, Bar Yam [3] recently offered a complex systems analysis of the recession and the Occupy movement. Other systems-theoretic frameworks that might be useful for understanding societal system pathologies are catastrophe theory [19], self-organized criticality [2], the adaptive cycle model [11], and scale-free networks [4].

Systemic problems require systemic solutions, i.e., structural change in societal differentiation. Major recessions, such as the 2008 crisis, and the inability of the political system of the United States to prevent them before they happen and counter them after they happen, point to basic flaws in the relations between economy, polity, community, and culture in the US societal system. Although concrete analysis of these flaws cannot avoid political judgments, seeing the problem as one of faulty differentiation gives this critique a non-combative tone. The diagnosis also suggests that with structural change these flaws are remediable. The organizing principle of a market economy is compatible with the organizing principle of democracy if—and only if—primacy is given to democracy.

Parsons' model does not reveal previously unknown societal problems or provide novel solutions to recognized problems. It is a framework within which known problems and possible solutions can be discussed coherently. Abstract models may be more effective than concrete models for identifying the essence or deep structure of societal problems, for seeing the forest rather than the many trees. From a complex systems perspective, the essence of the recession, viewed sufficiently abstractly, is in fact quite simple. Public understanding of the need for structural changes in the US societal system may be lacking currently, but an increase in such understanding in the future is imaginable.

The abstract model utilized in this chapter is "merely" graph-theoretical. That is, the model is mathematical in specifying a network of interactions or links, but the interactions are not themselves mathematically characterized in any way. Parsons' ideas could perhaps be given a more specific formalization. For example, the AGIL components might be defined as nominal variables having multiple states, and the decomposability of this tetrad—the degree of constraint between components—could be modeled quantitatively, even dynamically or diachronically, with Reconstructability Analysis [20]. This is an interesting research possibility for the future, but even a graph-theoretic model can provide useful perspectives on important phenomena in societal systems.

Acknowledgment

Thanks to Hugo du Coudray and Elinor Langer for helpful comments on this manuscript.

References

1. Adams R. N. (1975). *Energy and Structure: A Theory of Social Power*, University of Texas, Austin.
2. Bak P. (1996). *How Nature Works: The Science of Self-Organized Criticality*, Springer-Verlag, New York.
3. Bar-Yam Y. (2011). "Complex Systems and Occupy Wall Street," *http://necsi.edu/projects/yaneer/occupy.html.*
4. Barabási A. L., Albert R. (1999). "Emergence of Scaling in Random Networks," *Science*, **286**, 509–512.

5. Bateson G. (1979). *Mind and Nature: A Necessary Unity*, Bantam Books, New York.
6. Bennett J. G. (1961). *The Dramatic Universe: Vol. Two: The Foundations of Moral Philosophy*, Hodder & Stoughton, London.
7. von Bertalanffy L. (1979). *General System Theory*, George Braziller, New York.
8. Boulding K. (1978). *Ecodynamics*, Sage, Beverly Hills.
9. Gell-Mann M. (1994). *The Quark and the Jaguar: Adventures in the Simple and the Complex*, WH Freeman, New York.
10. Habermas J. (1984–1987). *The Theory of Communicative Action*, (McCarthy T, translator). Polity, Cambridge.
11. Holling C. S., Gunderson L. H. (eds.) (2002). *Panarchy: Understanding Transformations in Human and Natural Systems*, Island Press, Washington DC.
12. Jacobs J. (1994). *Systems of Survival: A Dialog on the Moral Foundations of Commerce and Politics*, Random House, New York.
13. Marx, K. (1977, first published1859) *A Contribution to the Critique of Political Economy*, Progress Publishers, Moscow.
14. Occupy Wall Street (2011). https://sites.google.com/site/the99percentdeclaration/.
15. Parsons T. (1966). *Societies: Evolutionary and Comparative Perspectives*, Prentice-Hall, Englewood Cliffs, pp. 5–29.
16. Parsons T. (1971). *The Systems of Modern Societies*, Englewood Cliffs New Jersey: Prentice-Hall.
17. Plato (1991). *The Republic: The Complete and Unabridged Jowett Translation*, Vintage Books, New York.
18. Simon H. A. (1962). "The Architecture of Complexity." *Proc. Am. Philos Soc*, **106**, 467–482. Reprinted in Simon (1981) *The Sciences of the Artificial*, MIT Press, Cambridge.
19. Zeeman E. C. (1977). *Catastrophe Theory: Selected Papers*, 1972–1977, Addison-Wesley, Reading.
20. Zwick M. (2001). Wholes and Parts in General Systems Methodology, in *The Character Concept in Evolutionary Biology* (ed. Wagner G.), Academic Press, New York. pp. 237–256.

Chapter 8

China's Complex Policy Network

Jerome Tan Sibayan

*The US Army War College, 122 Forbes Avenue, Carlisle Barracks, PA 17013, USA**

Kansas State University, Security Studies, 221 Eisenhower Hall, Manhattan, KS 66506, USA

jerome.t.sibayan.mil@mail.mil, jsibayan@k-state.edu

The Central Committee of the Communist Party of China is a complex policy network. The evolution of the network, whose relationships are defined by Chinese concepts of social connections known as *guanxi*, followed the predicted route to synchronization of scale-free networks. The network also exhibits complex characteristics of small worlds, emergent order, emergent functionality, and resilience.

8.1 Introduction

The Central Committee (CC) of the Communist Party of China (CPC) represents an influential political body in the world's most

*The views expressed in this chapter are those of the author and do not necessarily reflect the official policy or position of the Army, Department of Defense, or the US Government.

Complexity and the Human Experience: Modeling Complexity in the Humanities and Social Sciences
Edited by Paul A. Youngman and Mirsad Hadzikadic

ISBN 978-981-4463-26-3 (Hardcover), 978-981-4463-27-0 (eBook)
www.panstanford.com

populous country and the second largest economy. The CC ratifies fiscal policies and approves economic reforms, commits to military actions and modernization programs, and adopts security and foreign policies. Network analysis offers a way of revealing and illuminating the character and nature of the relationships between the CC members.

This study uses the Chinese conception of social relationships—*guanxi*, pronounced "gwahn-shee"—to determine which CC members have ties to other members. Later in this chapter, I will define the fundamental *guanxi* relationships and explain how the data is used. Network analysis software is used to analyze the basic data set for the Central Committee for each year from 1922 to 2011 to reveal network structures and patterns.

This chapter demonstrates how the CC's 90-year evolution closely followed scale-free network modeling predictions made by Yamir Moreno and others. In general terms, the CC's network began as a simple network with few members and few ties, but became a complex network of roughly two hundred full members exhibiting a distinct core-periphery structure. The Central Committee's structure and connectivity pattern reached a stable synchronized state in the last decade.

Next, this chapter shows how complex emergent order can be seen in the CC's network structure that is consistent with Robert Axelrod's agent-based models of competition, cooperation, and collaboration. His agent-based models predict regions of local convergence and global polarity in complex networks. The Central Committee's evolving network exhibits the emergence of regions of local convergence and global polarity.

Later, this chapter provides examples of complex emergent functionality in the Central Committee network at different time periods. Specifically, members of cohesive subgroups, or cliques, represent different functions of governance. This characteristic of complex systems is consistent with Stephen Wolfram's observation that simple abstract rules—in this case, rules of selection—can lead to complexity and emergence of differentiated functionality. Elite politics in China is well known for patron-leaders competing to promote client members of their own clique, especially when a new Central Committee is elected.

One important consequence of the selection and promotion of politically loyal members is the demotion and removal of members of other cliques. Frederick Teiwes believes that purges and rectifications represent behavioral norms of elite Chinese politics [1]. At times, so many CC members were purged that one wonders how the Party survived. This chapter argues that the Central Committee network is a resilient network—a robust complex body that does not collapse when random members retire, die, or are deliberately purged.

China scholar Lucian Pye notes, "Chinese political culture traditionally operated on the premise that omnipotence lies in the mystery which invisibility evokes" [2]. Network analysis is primarily a visual science and can shed light on patterns of relationships. This chapter helps to reveal the unseen nature of China's complex policy network by analyzing the network of relationships among members of the Central Committee.

8.2 China's Central Committee

The Central Committee is a very important policy network. Richard McGregor describes the Central Committee "as a kind of enlarged board of directors for the Party in China" [3]. Peter Moody characterizes the Central Committee as a "kind of parliament of the Party" [4]. The membership includes Party officials, government ministers, military officers, and provincial governors. Since 2002, the Central Committee officially includes entrepreneurs as members.

According to CPC statutes, the Central Committee selects the smaller and more powerful Political Bureau (Politburo) of roughly two dozen leaders from the CC membership. The Politburo selects the even smaller and more powerful Standing Committee of seven to nine leaders from among the Politburo membership. The Central Committee is fundamentally important in the selection of China's elite leadership and policymakers.

Although the Communist Party of China was established in the summer of 1921, the Party elected the "first" Central Committee in 1922. Since then until the early 1980s the election of CC members can be considered erratic at best. The Party elected the first five Central Committees almost annually. The next five were elected by the Party in 1945, 1956, 1969, 1970, and 1977. After Mao Zedong's death in 1976, the Party held regular Central Committee elections every five years.

As the Party membership grew from a few hundred in the 1920s to nearly eighty million today, the number of members in the Central Committees also increased. The first CC, confusingly named the "Second Central Committee" because it was convened after the Second National Party Congress, had only five full members and two alternate members. Nominally elected by the National Party Congresses, the lists of candidates for Central Committee membership are vetted and approved by the Party elite first. This simple abstract rule—a version of social homophily—ensures that new members have characteristics similar to the incumbent leaders and ultimately ensures the propagation of more general network characteristics.

With the exception of the political blood-letting of the first three years of the Cultural Revolution (1966–1969), the number of CC members steadily increased since it was established in 1922. However, since the era of Deng Xiaoping (1978–1992), the total number of full CC members has leveled off at roughly 200. The Eighteenth Central Committee of 2012 has 205 full members and 171 alternate members. Despite fluctuations in the number of members, the types of *guanxi* relationships between them are traditional in nature.

8.3 The Chinese Conception of Guanxi

Guanxi is ubiquitous in Chinese culture, society, economics, and politics. It is usually translated as relations, relationships, or connections. The cultivation and manipulation of *guanxi* plays a special role in Chinese life. Mayfair Yang's *Gifts, Favors, and Banquets* is the seminal scholarly contribution to the body of knowledge on *guanxi* [5]. She compares the Chinese to "spiders" weaving "magic webs" of *guanxi* networks. Cultivating relationships is something one has to do to obtain resources, get employment opportunities, and even garner political protection. The author interprets *guanxi* as a type of incipient relationship—that is, a gateway to further communication and opportunities for the convergence of interests. Lucian Pye argues that factional politics is based on clients establishing career-enhancing "linkages to power" based on which potential patrons they think will emerge victorious from leadership struggles [6]. In fact, power in Chinese politics typically equates status and prestige—oft-cited bases for webs of social (*guanxi*) relationships. The following sections present the four fundamental definitions of *guanxi*, which are used to define basic relational ties among members of the Central Committee's policy network.

8.3.1 Educational Level as Guanxi

The first type of relational tie is educational level. People who attend the same school have ties to each other. This is considered true even if they did not attend the same school or were students at the same time. Graduates sharing educational level, but different institutions, also have a social connection. Amy Hanser concludes that educational level (high school, technical college, and university) is a significant determinant of an individual's job prospects [7]. She finds that people are often linked according to similarities in education level. Such educational links enable people to achieve career success.

8.3.2 Military Experience as Guanxi

The second type of relational tie is military experience. Individuals who serve in the same unit or shared experiences (such as the Long March of 1934–1935) have a close relationship with each other. William Whitson and Chen-Hsai Huang provide the standard for defining military generations or cohorts: a group of men, regardless of age, who enter the military during a period that was characterized by a particular crisis [8]. For example, the Fourth Military Generation is the group that entered military service between January 1937 and December 1940. They shared the difficult process of recruiting and integrating new troops from North China. Officers belonging to the same military generations, regardless of educational level or provincial origin, are promoted as cohorts and therefore have relational ties to each other.

8.3.3 Provincial Origin as Guanxi

The third type of relational tie used in this study is provincial origin. Provincial boundaries often trace similar language dialects, ethics, values, and beliefs and serve as the basis for doing favors. Thomas Gold, Doug Guthrie, and David Wank consider ties based on "native place" (in addition to education, military experience, and kinship) important to defining the fabric of political culture and social life [9]. Shared geographic origin is easily seen as the basis for initiating contact with strangers regardless of culture. We often start conversations with "Where are you from?" Provincial is an obvious incipient relationship with people we do not know or are not related to. Patrons and kin, however, have inherent social relationships.

8.3.4 Special Connections

Special relationships fall into two categories: kin and non-kin. Claude Levi-Strauss examines how kinship forms alliance structures between European families [10]. Kinship is a "primordial" tie between people in that they exist prior to and are more important than all other kinds of relationships. Yunxiang Yan considers close kin as the "inner core" of social networks [11].

Two variations of kinship tie are patron-client relations and patrilineal bonds. Patron-client relationships are described as where an official controls underlings who depend on his political or bureaucratic power for protection and promotions. The patron-client relationship also requires reciprocity or a sense of obligation: patrons support clients' careers and clients support their patron's agenda.

Patrilineal bonds are parent-offspring relations. The ancient tradition of conveying or conferring power to a person on the basis of who his father is continues today. Central Committee members who are offspring of well-known Party leaders are called "Princelings." Some China observers note that the Princelings earlier held some level of power parity with the other established political factions during the Hu Jintao era (2002–2012). The Politburo and Standing Committee elected in late-2012 are dominated by Princelings.

All four fundamental guanxi relationships are used in this study to define the Central Committee's network structure. Ties based on educational levels, provincial origins, military experience, and special connections pervade Chinese society and politics. Fortunately, basic biographical information about Central Committee members is readily available for network analysis.

8.4 Recreating the Network

8.4.1 Sources, Interpretation, and Coding of Data

This study uses two sources of biographical data of each member of the Central Committee since 1921. Wolfgang Bartke's *Biographical Dictionary and Analysis of China's Party Leadership, 1922–1988* provides the majority of information needed to determine *guanxi* similarities [12]. The other source is the Party History Research of the Organization Department of the Communist Party of China's *Past Chinese Communist Party Central Committee Dictionary, 1921–2003* [13]. These volumes provide education levels, military assignment history, provincial origin, and mention any special relationships. The network analysis that follows uses other histories of the Communist Party of China to cross reference information wherever possible.

First, this study identified members of each Central Committee. The analysis required revising and updating the list of members for each year from 1922 to 2011 based on purges, retirements, and deaths. On occasion, the Party elected alternate CC members to full membership to fill vacancies. As part of the analysis, I collected basic data about individual attributes of each Central Committee member. The data notes which CC members were also Politburo members, Standing Committee members, Long Marchers, technocrats, and military officers. These additional attributes add context to the standard network data already gathered.

The analysis coded biographical data (members' names, their relationships, and individual attributes) for analysis using network analysis software [14]. The network analysis software analyzed the data for individual centralities, group centralization, and network graphs and generated structural measurements for each Central Committee for each year. At this point, you are invited to watch a short 3-minute movie that demonstrates how the Central Committee's network structure has changed over time. (Go to http://www.youtube.com/watch?v=j_ZG-xEMMEU&feature=youtu.be.)

8.5 Five Characteristics of a Complex System

There are potentially many criteria a system must meet in order to be called "complex." Scott Page argues that complex systems are moderately interdependent, moderately connected, somewhat diverse, and somewhat adaptive [15]. This section suggests that complex systems exhibit scale-free self-similarity, small-world characteristics, emergent order, emergent functionality, and resilience. One sees these observable characteristics in the Central Committee's social network.

Benoit Mandelbrot's fractal geometric models are some of the best-known examples of scale-free self-similarity [16]. Regardless of whether the observation is made at the macro or the micro level, the patterns in scale-free networks are the same. Steven Strogatz opines that "social networks also seemed likely to be small worlds," that is, small and highly clustered [17]. Emergent order suggests that simple rules applied locally (i.e., at the bottom) lead to macro-level properties. Emergent functionality suggests that patterns of functional diversification will emerge within originally homogeneous systems. (Scott Page calls emergent functionality: "breaking symmetry"). Resilience suggests that complex networks continue to serve their intended purpose despite the random loss of members of their network communities.

8.5.1 A Scale-Free Network

Yamir Moreno suggests that scale-free networks evolve along a particular route or pathway to synchronization or stability [18]. The images in Fig. 8.1 are from Moreno's model and represent the different phases a scale-free network is expected to follow as it achieves a stable state. The first image on the left is of an initial-state network with low coupling intensity resulting in a poorly defined core surrounded by mostly disconnected clusters. In the second image from the left the core is growing, but is still poorly defined. The third image suggests that the entire network, although altogether more connected, lacks organization and cohesion. The last image, intended to represent a stable state of synchronization, clearly shows the evolution of a highly cohesive core surrounded by a thin periphery connected to the core.

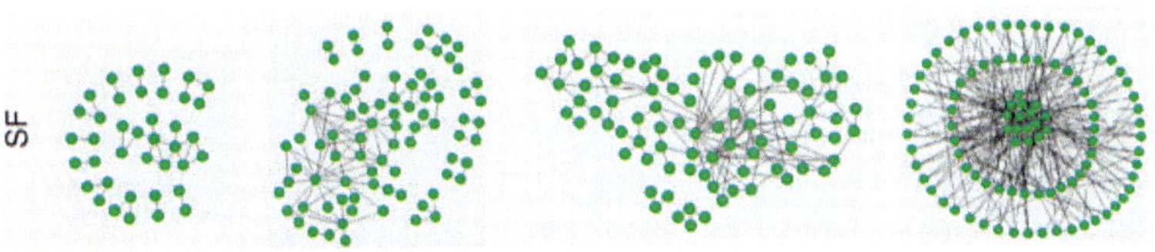

Figure 8.1 Y. Moreno's model of the route to synchronization of a scale-free topology [18].

Compare Moreno's computer model of the evolution of a scale-free network (Fig. 8.1) with the actual evolution of China's Central Committee (Fig. 8.2) taken from the network analysis for this study. The similarity between the two sets of images suggests that the evolution of China's Central Committee bears a striking resemblance to Moreno's prediction for scale-free networks.

Figure 8.2 The Central Committee's route to synchronization.

By definition, scale-free networks have similar patterns at different levels of analyses. Therefore, we should expect to see the same distinct core-periphery pattern at *every* level of the Chinese political system. Unfortunately, space limitations prevent further analysis here. Nevertheless, the Central Committee's social network diagrams above suggest that the CC is, in one important sense, a self-contained world.

8.5.2 A Small World After All?

Duncan Watts suggests that networks exhibiting small world phenomena meet four criteria [19]. First, they are numerically large networks consisting of much more than one member. Second, the networks are "sparse" in the sense that members are connected to an average number of other members much smaller than the total number of members. Third, the number of connections enjoyed by the member with the most connections is also much smaller than the total number of members. Finally, such networks are highly "clustered," meaning that those with links to a particular member are also linked to each other. Specifically, small world networks have relatively high cluster coefficients—approaching 1.0 and they also have relatively low average path lengths or distances among all reachable members.

In the graph shown in Fig. 8.3, it is apparent that as the Central Committee's membership increased over the years, its cluster coefficient has steadily risen to nearly 0.9 (blue line and left scale). Similarly, the average distance of the network has varied only slightly around 2.0 (red line and right scale). It is noteworthy that the cluster coefficient and the average distance have approached 1.0 and 1.5, respectively, since the end of the Red Guard Phase of the Cultural Revolution, when mass purges last occurred, and at the same time that the number of full Central Committee members asymptotically approached 250. This appears to be consistent with Duncan Watts' predictions for small-world, agent-based networks.

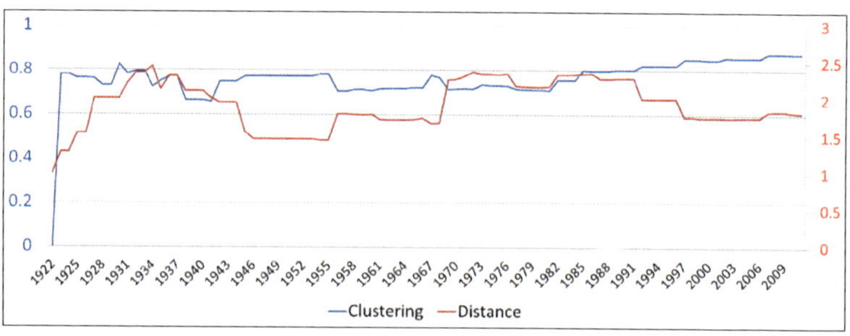

Figure 8.3 Change over time of the Central Committee's clustering coefficient and average distance.

8.5.3 Emergent Order

Robert Axelrod's work on agent-based models of competition, cooperation, and collaboration demonstrates that certain end states are typical of complex interactive systems [20]. His models of cultural dissemination suggest that communities of members with somewhat differentiated domains, or characteristics, whose interactions are rule-based, will eventually arrive at an equilibrium state. Specifically, in the end there will be a number of stable, but differentiated, regions within which different characteristics are dominant. Axelrod argues that local convergence (or clustering) leads to global polarization.

A closer look at how the CC's network diagram evolved since 1981 is suggestive of Axelrod's prediction of a trend toward equilibrium. In 1981, the Central Committee had 23 k-cores (Fig. 8.4). There were 194 full members of the Central Committee that year and each member had an average of 28 connections. After nearly sixty years of existence, the policy network of 1981 exhibited two or three main cohesive subgroups with several smaller cohesive subgroups on the periphery. Axelrod might describe the network by noting "the emergence of distinct cultural regions" defined as contiguous areas with similar types of connections. The network diagram of the Central Committee in 1981 (Fig. 8.4) is similar enough to a computer model of Axelrod's "dissemination of culture" (Fig. 8.5) [21] to warrant comparison on the basis of identifiable subgroups.

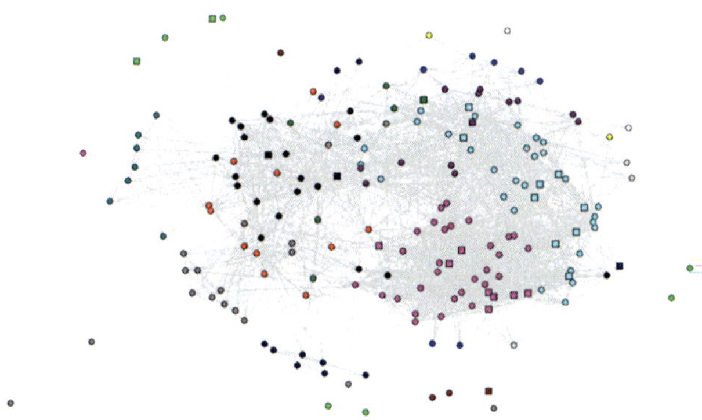

Figure 8.4 The Central Committee of 1981.

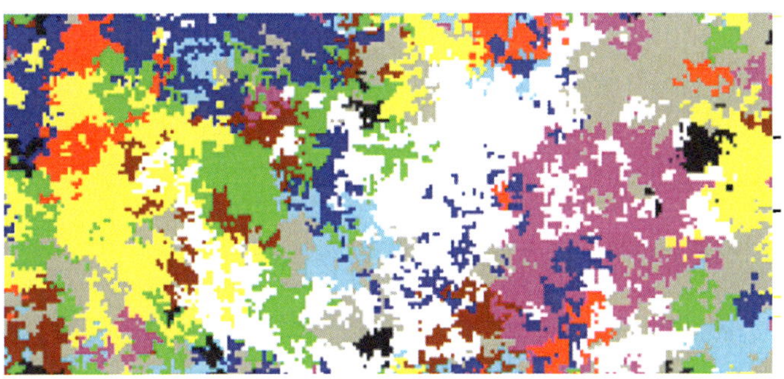

Figure 8.5 Computer model of Axelrod's "dissemination of culture" [21].

A look at the Central Committee network diagram of 2007 with 15 k-cores makes the point that over time the number of coherent regions or cohesive subgroups has emerged (Fig. 8.6). The reduction in the number of k-cores from 23 to 15 is consistent with Axelrod's prediction that the number of regions will asymptotically approach a stable number of coherent regions. In the 2007 network diagram, one k-core clearly dominates the central position of the network, while the other 14 k-cores reside along the periphery. The network's year-to-year metamorphosis from 23 to 15 k-cores cannot be shown here, but it is apparent that at least one k-core, in becoming dominant, has subsumed the members of the other k-cores, which consequently became minor cohesive subgroups. Complex networks are not only scale-free, small-world, and increasingly ordered, but they are also functionally differentiated.

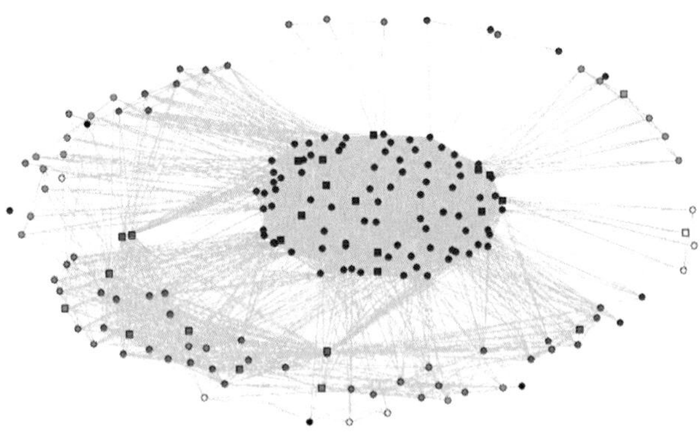

Figure 8.6 The Central Committee of 2007.

8.5.4 Emergent Functionality

Kenneth Lieberthal observes that political factions "place at least one of its supporters into each of the major functional hierarchies" [22]. Turning this around a bit, we can alternatively say that major functional hierarchies are represented in each of the various factions. In order for such a strategy to be effective for a faction, and assuming that *most* Party members do not hold more than one position in different functional bureaucracies, a controlling faction must be sufficiently large in number to enable the faction members to influence *most* of the Party and state functions. Two examples suggest that functional differentiation may be detected as either the type of relations among members or, as Lieberthal suggests, as factional representation in the major functions of government.

An analysis of the Central Committee network for 1942 demonstrates the emergence of functionality in terms of types of connections and in terms of the coalescing of functional domains. The colors indicate different types of connections (Fig. 8.7). Green lines represent ties between CC members who share the same level of education. Red lines indicate people coming from the same province. Blue lines show connections between members belonging to the same military cohort. The patterns of connections are readily apparent. The cohesive subgroup on the far left is primarily bound together by provincial ties. The cohesive subgroup on the far right is linked by educational ties. The two groups are connected by similarities in military experience.

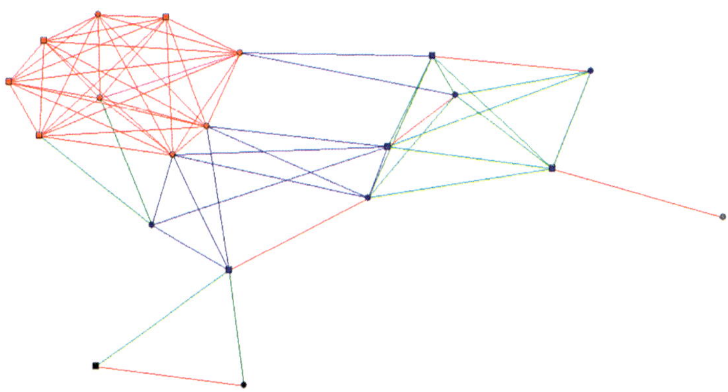

Figure 8.7 The Central Committee of 1942.

A closer inspection of the group on the right (Fig. 8.7) bound by educational connections includes Zhou Enlai, who was the Secretary of the Central Committee's Secretariat at the time. Nearby in the network was Kang Sheng, Director the CC's Information Department; Wu Yuzhang, President of Yan'an University; Wang Jiaxiang, Director of the 8th Route Army Political Department; and Deng Fa, President of the Central Party School. Training, indoctrination, and education are common functional threads among their disparate places within the Party-state structure.

The group on the left (Fig. 8.7) connected by shared provincial origins also shares functional purposes among its members. In 1942, Mao was a member of the Central Revolutionary Military Council and still a year away from being named Party Chairman. In the same cohesive subgroup was Peng Dehaui, Deputy Military Commander of the 8th Route Army; and Liu Shaoqi, Political Commissar of the New 4th Army. Clearly, they shared military operational responsibilities during the war against Japan, on which the Central Committee was focused in 1942.

The number of Central Committee members doubled from 23 in 1942 to 44 in 1945, when the Seventh Central Committee was elected. Once again, the pattern of connection types is readily apparent (Fig. 8.8). This time the core group in the middle is bound primarily by military cohort ties (blue lines). The upper half of the network is tied together by shared provincial origins (red lines). And two groups on the left and right side are connected by similar educational backgrounds (green lines).

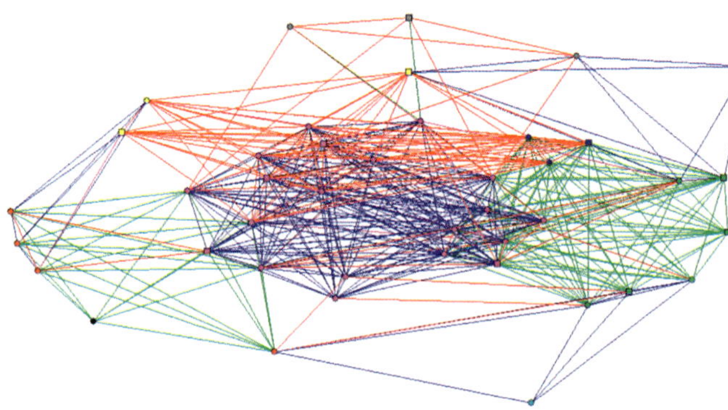

Figure 8.8 The Central Committee of 1945.

Kenneth Lieberthal describes China's political system as one of "fragmented authoritarianism," that is, authority below the very peak of the Chinese political system is fragmented and disjointed [23]. More specifically, the bureaucratic ranking system combines with the functional division of authority among an array of bodies. Authority is decentralized throughout different localities and bureaucratic units.

It should be noted that functionalism emerges within the various cliques and factions. For instance, the six-member clique that formed on the right-hand side of the 1945 network (Fig. 8.8) includes one, and only one, member with functional responsibilities in the North China Military Region (Deng Xiaoping). The same is true for the Northeast China Military Region (Gao Gang), the Central Military Region (Deng Zihui), and the New Fourth Army (Rao Shushi). One clique member (Kang Sheng) is assigned to the Central Committee's Organization Department and one is assigned to the Central Party School. There is virtually no redundancy of responsibilities, resulting in an extremely efficient clique in which each clique member is solely responsible for a government function.

Another example of functional differentiation within cliques is the clique that formed at the upper portion of the 1945 network (Fig. 8.8). This six-member grouping, connected by shared provincial origin, includes one, and only one, member with functional responsibilities in the North China Bureau (Dong Biwu), the Central Military Region (Li Xiannian), the All-China Federation of Labor (Liu Shaoqi), the Central Committee's Organization Department (Li Fuchun), and the People's Revolutionary Military Council (Mao Zedong). The similarity between the differentiated reach of the two cliques just described is more than coincidental.

I expected to find one clique dominated by the Ministry of Defense and another dominated by members assigned to the Ministry of Foreign Affairs—each cohesive subgroup representing a particular function. In other words, I hoped to find that members of the same cohesive subgroup held similar functional responsibilities, which would be consistent with an earlier study of elite decision makers [24]. Instead, each cohesive subgroup is functionally differentiated *at its own level*. In other words, differentiated functionality emerged within each clique. This is very notable for a couple reasons. First, each peripheral clique, perhaps aspiring to become the core clique, is efficient in that there is no redundant coverage of particular state or governmental functions. Second, in order to become a credible clique capable of assuming legitimate control of critical government functions, the "tentacles" of each clique must reach down to the most important provinces and into the most important ministries. A peripheral clique is competitive only if it is both functionally representative and potentially pervasive.

8.5.5 Resilience of Guanxi Networks

Mayfair Yang argues that *guanxi* should be treated as a multi-faceted ever-changing phenomenon. *Guanxi* networks persist in the "fluctuating streams of history" and adapt to new institutional arrangements because the practice of and need for *guanxi* persists. The Central Committee is a resilient policymaking network that is based on *guanxi*. The Central Committee network is an organic system that is fluid and flexible, within which, as discussed earlier, individuals perform a variety of tasks.

In order for the entire Central Committee to be a resilient network, each function must have a fail-safe or backup system. The loss of a single person (or a randomly selected group of people) must not prevent the entire organization from fulfilling its intended purpose. As demonstrated earlier, different cliques held functional positions in the same locality or bureaucratic body. In other words, the elimination of a single clique did not result in the elimination of the Central Committee's authoritarian grip on geographical or governmental organs of power. The Central Committee's constitutional purpose is to serve as the "supreme power to govern Party affairs and to enact Party policies ... Party and state policies and programs are discussed and ratified" [25].

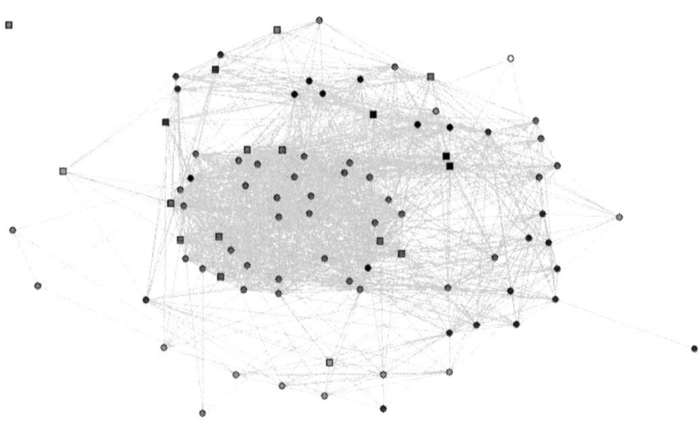

Figure 8.9 The Central Committee of 1965.

Perhaps the best time period to examine the Central Committee for evidence of resiliency lies with the Great Proletarian Cultural Revolution (1966–1976). The network diagram of the Central

Committee just prior to the Cultural Revolution is shown in Fig. 8.9. The major Party organizations experienced purge rates of 60% to 70%. Nine of the twenty-three Politburo members survived. Only 54 of the 167 Central Committee members' political careers remained intact.

In 1966, Mao purged thirteen Central Committee members including Deng Xiaoping (Party Secretary General), Lu Dingyi (Propaganda Department), Luo Ruiqing (PLA Chief), Bo Yibo (State Planning and Economic Commissions), and Peng Zhen (Mayor of Beijing). Nevertheless, the Party and the State continued to function. Although the purged leaders did not all belong to the same faction or clique, they were accused of being "revisionists." Their functional replacements all came from within the Central Committee.

Harry Harding observes that the purge rates were not uniform throughout the state bureaucracy [26]. Functional areas such as agriculture, industry, planning, culture, and education suffered high purge rates. The network diagram of the Central Committee in 1969 is shown in Fig. 8.10. National defense, finance and trade fared much better. Through it all the People's Republic of China managed to raise grain production, relocate heavy industry to the country's interior, go toe-to-toe militarily with the Soviet Union, and begin rapprochement with the United States and the West [27]. Despite the chaotic excision of Central Committee members, the Party not only remained in control, but also effectively pursued a variety of policies.

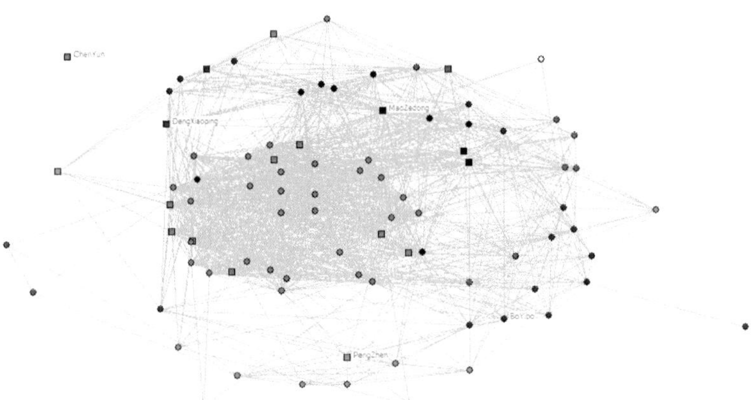

Figure 8.10 The Central Committee of 1969.

Taking a different approach to demonstrating resiliency, ten percent of the members of the Central Committee of 2011 were randomly removed by deleting every tenth member from an alphabetical listing of the membership (Fig. 8.11). Comparing the original, actual network (on top) with the randomly purged network (on bottom), it is readily apparent that the network structure is robust in the sense that it did not substantively change. In other words the pattern of relationships and the array of cliques were not affected by randomly purging individuals and it is therefore plausible to expect that the network will continue to function as it did before the purge.

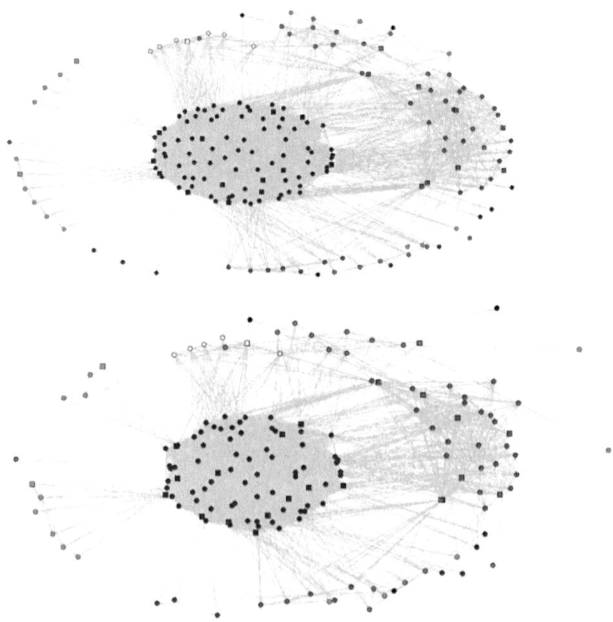

Figure 8.11 The Central Committee of 2011. The upper diagram is the actual network, while the lower diagram is the randomly purged network.

8.6 What a Complex Policy Network May Mean

This chapter argues that the Central Committee of China's Communist Party is a complex policy network that is scale-free, characteristically small-world, becoming ordered, functionally differentiated, and resilient in the face of random disturbances. The fact that the Central Committee is scale-free suggests that the network operating conditions occurring at the Central Committee level are also operating at all other levels of the Party's organization: provincial, municipal, and county. Further, the more "mature" a policy network is, the more cooperative are its internal and external policies. Therefore, it is desirable to allow scale-free structures to evolve and develop normally toward stable states of synchronization.

The idea that China's Central Committee is becoming more ordered, less fragmented, and more homogeneous suggests that the Central Committee is also becoming more polarized. Is this leading to the formalization of a two-party system? Probably not any time soon, but the potential is certainly there for the development of open divisions along two primary lines of policies. There have always been Leftists and Rightists, Revolutionaries and Revisionists in China's political history. Perhaps the incremental ordering of China's Central Committee is indicative of the development of opposing political parties, but it is almost certainly pluralism within of the Communist Party.

Finally, the most profound quality of the Central Committee as a complex policy network is its resiliency. Given the absence of a single, strong dictator leading the Communist Party, the era of violent (even murderous) purges are probably long gone. Nevertheless, should we ever witness mass purges of the Party leadership in the future; we can expect the Central Committee to adequately function as a policy ratifying and policy implementing network—all else being equal.

References

1. Teiwes, F. (1979). *Politics & Purges in China: Rectification and the Decline of Party Norms, 1950–1965* (M. E. Sharpe, New York).
2. Pye, L. (1995). "An Introductory Profile: Deng Xiaoping and China's Political Culture." In *Deng Xiaoping, Portrait of a Chinese Statesman* (ed. Shambaugh, D.), Clarendon Press, New York.
3. McGregor, R. (2010). *The Party: The Secret World of China's Communist Rulers* (Harper, New York).
4. Moody, P. (1971). *The Politics of the Eighth Central Committee of the Communist Party of China* (Shoe String Press, Hamden).

5. Yang, M. M. (1994). *Gifts, Favors, and Banquets: The Art of Social Relationships in China* (Cornell University Press, Ithaca).
6. Pye, L. (1981). *The Dynamics of Chinese Politics* (Oelgeschlager, Gunn & Hain Publishers, Inc., Cambridge).
7. Hanser, A. (2002). "Youth Job Searches in Urban China: The Use of Social Connections in a Changing Labor Market." In *Social Connections in China: Institutions, Culture and the Changing Nature of Guanxi* (ed. Gold, T., Guthrie, D., and Wank, D.), Cambridge University Press, New York.
8. Whitson, W., and Huang, C. (1973). *The Chinese High Command: A History of Communist Military Politics, 1927–1971* (Praeger Publishers, New York).
9. Gold, T., Guthrie, D., and Wank, D. (2002) "The Study of Guanxi." In *Social Connections in China: Institutions, Culture and the Changing Nature of Guanxi* (ed. Gold, Guthrie, and Wank), Cambridge University Press, New York.
10. Levi-Strauss, C. (1969). *The Elementary Structures of Kinship* (Beacon Press, Boston).
11. Yan, Y. (1992). "The Impact of Rural Reform on Economic and Social Stratification in a Chinese Village." *Aust J Chin Aff*, **27**, 1–23.
12. Bartke, W. (1990). *Biographical Dictionary and Analysis of China's Party Leadership, 1922–1988* (K. G. Saur, New York).
13. Party History Research of the Organization Department of the Chinese Communist Party. (2004). *Past Chinese Communist Party Central Committee Dictionary, 1921–2003* (中国共产党历届中央委员大辞典, 1921–2003) (Beijing).
14. Borgatti, S. P., Everett, M. G., and Freeman, L. C. (2002). *UCINET 6 for Windows Software for Social Network Analysis* (www.analytictech.com, accessed June 2010).
15. Page, S. (2009). *Understanding Complexity (lectures)* (The Teaching Company, Chantilly, VA).
16. Mandelbrot, B. B. (1977). *The Fractal Geometry of Nature* (W. H. Freeman and Company, New York).
17. Strogatz, S. (2003). Sync: *How Order Emerges from Chaos in the Universe, Nature, and Daily Life* (Hyperion, New York).
18. Barrat, A., Barthelemy, M., and Vespignani, A. (2011). *Dynamical Processes on Complex Networks* (Cambridge University Press, New York).

19. Watts, D. J. (2003). *Six Degrees: The Science of a Connected Age* (Norton, New York).

20. Axelrod, R. (1997). *The Complexity of Cooperation* (Princeton University Press, Princeton).

21. Leydesdorf, L. (2001) "Technology and Culture: The Dissemination and the Potential 'Lock-in' of New Cultures." *J Artif Soc Soc Simulation,* **4**(3) (http://jasss.soc.surrey.ac.uk/4/3/5.html, Accessed July 10, 2012).

22. Lieberthal, K. (1978). *Central Documents and Politburo Politics in China.* Number 33 (Michigan Papers in Chinese Studies, Ann Arbor).

23. Lieberthal, K. (2004). *Governing China: from Revolution through Reform* (W. W. Norton & Company, New York).

24. Alba, R. D., and Moore, G. (1978). "Elite Social Circles." *Sociol Methods Res,* **5**, 167–188.

25. Dreyer, J. T. (2012). *China's Political System: Modernization and Tradition* 8th ed. (Longman Pearson, New York).

26. Harding, H. (2011). The Chinese State in Crisis, 1966–1969. In *The Politics of China: Sixty Years of the People's Republic of China* (Cambridge University Press, New York).

27. Lieberthal, K. (1992). *Bureaucracy, Politics, and Decision Making in Post-Mao China* (University of California Press, Berkeley).

Chapter 9

Convergence versus Emergence of Youth Extremism: An Agent-Based Model of the Arab Spring

Dana Downey

George Mason University, School of Public Policy,
3351 Fairfax Dr, Arlington, VA 22201 USA

danadowney@gmail.com

This chapter examines whether patterns of convergence and emergence explain radicalization in an epistemic network of the Arab Spring. A hypothetical Islamic network is coded in NetLogo to create patterns of spatial grouping determined by agent adaptability to their peers in a Moore neighborhood. Influence rate and employment determine the spatial distribution of the agents. Convergence rather than emergence is revealed as the driver of network clustering and hence explains radicalization among the majority of the agent subset.

9.1 Introduction

The events marked by the first political demonstration in Sidi Bouzid in Tunisia on December 18, 2010, following the self-immolation by Mohamed Bouazizi provide social science an opportunity for

Complexity and the Human Experience: Modeling Complexity in the Humanities and Social Sciences
Edited by Paul A. Youngman and Mirsad Hadzikadic

ISBN 978-981-4463-26-3 (Hardcover), 978-981-4463-27-0 (eBook)
www.panstanford.com

testing the correlation between tipping points and social protests. The so-called Arab Spring opens the door for examining the links between tipping points and contextual drivers. A tipping point is a type of change in a system's dynamic that leads to large-scale transformation. Contextual tipping points are a series of micro changes that cause a shift in the macro outcome. In this agent-based model, I measure the effect of unemployment on agent actors and suggest why unemployment, especially among the youth population, was a key driver that triggered a tipping point in the Arab Spring.[a]

This chapter attempts to explain the behavior in social networks during the Arab Spring that caused the breakdown of two long-term tyrannies in the Middle East, that of Zine el Abidine Ben Ali on January 14, 2011, and Hosni Mubarak on February 11, 2011. This agent-based model is designed to capture the dynamics of how groups function as collective vehicles of belief propagation, and to reveal the conditions under which informal substructure can change from being system supporting to being system destabilizing, and what the relationship is between belief diffusion and social insurgency. In Sections 9.4 and 9.4.1, I reference existing social science theories of collective action that are applicable to describe the social context of the Arab Spring.

Existing theories in computational dynamics do not hold in the simulated environment that I describe in this chapter as far as telling us what led to the political transitions in the Arab Spring [2,4,5,10,12,17]. Some of these theories state that sight radius is a factor [12]. Sight radius in computational dynamics describes the distance between agents in a computational population set. I offer an alternative explanation for a mechanism that causes social unrest to reach a tipping point that leads to political change in society.

The 2011 uprisings in the Middle East demonstrated that there were many factors in society that were on the verge of tipping into unrest for many years. One of these primary factors is the so-called

[a]The term "Arab Spring" appeared in media to describe the uprisings in Tunisia. On January 4, 2011, *Christian Science Monitor* asked whether the Middle East was experiencing an "Arab Spring?, Or Arab winter?". Again on January 6, 2011, Marc Lynch of *Foreign Policy* inquired, "Are we seeing the beginnings of the Obama administration equivalent of the 2005 "Arab Spring." A US commentator first used "Arab Spring" after a short-lived Middle Eastern democracy movement in 2005. The term is also connected to George W. Bush, who referenced the phrase to refer to Arab leaders who were opening political spaces for citizen activism in 2005 [18].

youth bulge. The youth imbalance had a profound effect on the events in Egypt's political uprising when understood in the context of military control, government policy, socio-economic inequality, and elite overproduction. This results when the number of seekers exceeds the limited supply of elite positions, increasing both rivalry and factionalism within the elite. In the case of Egypt, this led to an expansion of the army and the bureaucracy in a constrained economy that did not offer expanded taxation to cover rising real costs.

9.2 The Effect of the Youth Bulge on Collective Action

The Middle East is experiencing one of the largest youth bulges in the world. Thirty-three percent of Middle Eastern society is under the age of 15. Egypt is typical. Fifty-four percent of Egypt's population is under 24 years of age and the ratio of young people in Egypt (ages 18–29) is nearing 19 million or 24.2% of Egypt's population of 81 million [8,21].

The growing number of unemployed youth created a security concern that the Mubarak regime was unable to address and it became a persistent source of grievances in the years that led up to the protests. Government policy planning under Mubarak was unable to reduce the pressure of the growing youth population on the economy. Youth unemployment soured in part because of pressure from the international financial community, which impelled Mubarak to reduce parastatal and other subsidized sources of employment. The state sector was trimmed to reduce government deficits. Unfortunately, the economy did not grow fast enough to create new goods to address the demands of a rising population. While the regime tried to stimulate the private sector, investment corruption, red tape and complex regulations deterred investment. As a result, policies of state entrenchment did not create sufficient economic opportunity to absorb the loss of employment and prevent the political mobilization of the unemployed youth in Egypt. Youth services were never the focal point of national development policy. The thorniest problem was that the education system was turning out a large number of degree holders who had limited practical job opportunities.

Mubarak was cheered by the international financial community for his efforts to reduce the burden of the state sector on the government budget. In fact, Egypt received acclaim from the World Bank for having done more to improve its business sector than any country in the world, which enabled the Egyptian government to mount international investment capital for over seven years, but its effort to create a conducive environment for private investment was not sufficient to make up for the losses of employment and the trimming of the parastatal sector. The World Bank ranked Egypt as one of the top 10 performers out of 185 countries assessed, four times, two years successively, between 2002 and 2009, for making more progress than other countries toward improving the regulation of the business environment. The bank noted, in their annual "Doing Business" report, Egypt's progress toward simplifying the startup process, gaining access to credit, contract enforcement and eliminating red tape as reform highlights [6]. However, the World Bank did not recognize that state contraction increased the vulnerability of the regime to the disappointments caused by shortage of elite populations. This affected the youth because there were fewer opportunities to obtain positions that conferred social status.

In both Tunisia and Egypt, middle-class unemployed youth made up the bulk of the protesters. In Tunisia, total unemployment rose to 16% in 2010 and in Egypt to 9.7% that same year. According to official government figures this ratio in Egypt was further exacerbated in the 18–29 age group, for which the rate of unemployed youth reached 20.4% in 2010 [21].

> The ratio was higher for males with college degrees than for holders of high school diplomas and holders of technical education diplomas, at 26.8 percent, 19.3 percent and 11.2 percent, respectively. The ratio was particularly high among females with a college degree and females with high school diplomas, at 56.1 percent, 55.1 percent, and 45.5 percent, respectively. The ratios were even higher in the survey of young people conducted in 2009, which showed that nearly 58.5 percent of youths in the above age group were unemployed; the ratio being 83 percent for females and 27 percent for males [21: 40].

The unemployed were the strongest force in the protests. Unemployment among Egyptian college graduates was 10 times higher than those without a degree [8]. The only Egyptians who did not participate in the riots were the wealthy elites associated with the regime along with the very poor and impoverished peasant classes [3: 67]. Unemployed Tunisian members of the middle class also represented a wide subset of the population involved in the riots [3].

Islamic parties and as well other civilian organizations in both nations played a role in inciting public contempt for the regime months before the protests erupted in 2011. Many of the unemployed spent their time outside of the home participating in societal organizations and social movements, to which the youth turned for mentorship and education. Their identity became strongly connected to such movements since they did not have an identity as wage earners and they spent little time in economically fruitful activities. "These groups offered new channels for youth activists because of their clear goals, proper understanding of youth problems, and willingness to try new methods of action, all of which contrasted with the government's outmoded ways" [21: 44]. Numerous activist organizations in Egypt—The April 6 Youth Movement, We Are All Khaled Said, the National Association for Change, Kefaya (Egyptian Movement for Change), and the Muslim Brotherhood—were all key contributors to the uprisings. In Tunisia, it was the Union Générale des Travailleurs Tunisiens (UGTT), the nation's strongest labor union that helped organize the demonstrations due to the fact that their membership was composed of civilians who were experienced in the art of political protests [7].

9.3 Modeling the Youth Bulge in an Agent Population

This chapter offers a model to explore the influence of the workplace and social environments on belief diffusion among an agent subset. My sample population is based on a hypothetical Islamic environment, in which I create a scenario to resemble the real-time daily routine in Egypt prior to the revolution. In this agent-based model (ABM) the marker of "fundamentalist" in the agent subset represents unemployed youth. I assess whether routine employment would control extremist activity by limiting the number of hours, per day, each employed agent spends in a work environment, in the sense that introducing a positive social structure makes a difference on the mentality behavior of agents. To do this, I populated an ABM with an agent set that responds to a randomized numeric algorithm maneuvers the agents over four quadrant zones. Each quadrant zone determines with whom an agent interacts; furthermore, it establishes the time frame of each interaction. For example, employment as an independent variable might restrict an agent's mobility and interactivity during the day and consequently its likelihood to influence other agents or be influenced by agents and contribute to the overarching fundamental level in the model. Thus, this work attempts to understand how institutions might affect populations in the same way peer groups and value systems seem to. The agent population is instructed to distribute spatially (through the cellular automata in the model) and interact within its Moore neighborhood. Moore neighborhoods are a set of cells surrounding a given, center cell. Moore neighborhoods give an agent geographical proximity to another [5]. Monitoring the Moore neighborhood is the process of calculating the agent's behavior in any given patch; the agent in the patch reacts to its surrounding neighbors, which includes itself and the four adjacent sides of the one center patch and the patches touching those corners (a Moore neighborhood has eight total sight areas) (Fig. 9.1). For instance, inside one patch an agent can be considered as having a "sight radius" of one [12].

The Moore neighborhood in this model is used as a parameter space for which agents influence one another. In earlier research, Lustick and Miodownik find that "the smaller the sight radius of agents in a landscape, the more 'local' are the interactions driving the behavior of individual agents and therefore driving the behavior of the array as a whole" [12]. This type of behavior was reported in the work of American sociologists Herbert Blumer and Robert E. Park. Blumer and Park noted "that members of most crowds or temporary gatherings engage in individual sequences of behavior and occasionally in sequences of behavior in common or in concert with many other members of the gathering. However, the form of social behavior in which they most frequently engage is some form of interaction, including conversations, within their small groups of family, friends or acquaintances" [20: 387,13: 415]. It can also be said that as the number of individuals in any gathering increases, the possibility could increase for every individual to engage in conversation with other individuals in that gathering [13: 416].

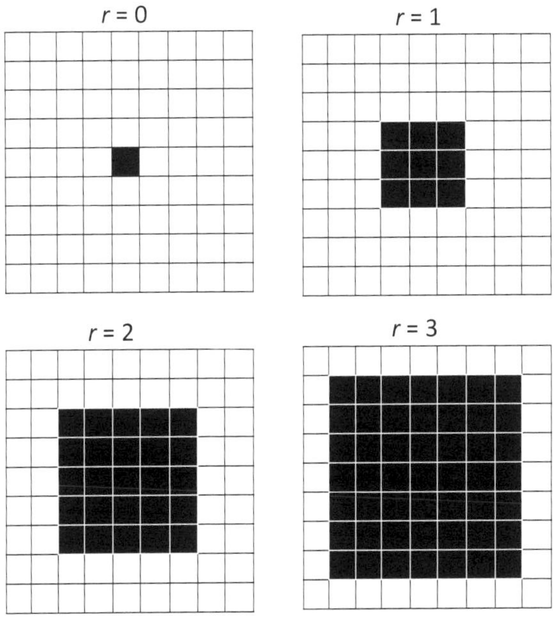

Figure 9.1 Four examples of a Moore neighborhood, the sight radius for each corresponds to *r* = 0 SR0, *r* = 1 SR1, *r* = 2 SR2, *r* = 3 SR3 [22].

The spatial parameter that I monitor for group influence is measured by a normal age distribution of a population and based on the likelihood of an agent's susceptibility to peer group pressure during certain stages of its life, thus, the influence rate measures the later. The influence rate varies as shown in Table 9.1.[b]

[b]The age spread is based on a cumulative distribution in a normal curve. The influence rate for each age group is: 12–20 (rate of influence is multiplied by 12); 21–30 (rate of influence is multiplied by 8); 31–40 (rate of influence is multiplied by 2); 41–50 (rate of influence is multiplied by 6); 51–60 (rate of influence is multiplied by 4); and 61–72 (rate of influence is multiplied by 5).

Table 9.1 The influence rate for each age group ages (12–72) and rate (0.00–1.0)

Age	Rate
12–20	100% chance of influence
21–30	66% chance of influence
31–40	15% chance of influence
41–50	50% chance of influence
51–60	33% chance of influence
61–72	45% chance of influence

The environmental context in this model is a graphed landscape, in which agents operate based on stylized assumptions of the role of daily culture in the Middle East and/or Islamic states. The agents' spatial distribution is determined by a random number generator designed to generate a distribution sequence for the agents in a randomized pattern, and disperses the population over four quadrants—home, mosque, work, and market. This randomized pattern is based off a random number seed generator in NetLogo software. It is a pseudo-random number generator as opposed to a true random number generator. It is configured by a deterministic equation and computes an algorithmic variable based on a predetermined seed variable. Agents are instructed during the startup function to initiate in the home quadrant and from there transverse the landscape as directed by the random seed generator that determines their gender and employment position (Fig. 9.2). The model starts all agents in the home quadrant and begins at 1:00 am, based on a 24 h time scale. Agents then leave home 7 h later and proceed as algorithmically instructed for the remainder of the 24 h time circuit. To control for specific regularity at the agent level and to accomplish the goals of creating a model that replicates a Middle Eastern norm, each agent is instructed to pass through and spend 2 h at the mosque and 6 h at the market each day. The remaining 9 h an agent's spatial distribution is activated by the random number generator that responds by providing an agent's age and employment position. Such control factors are manipulated by the global variables that are located in the interface of the NetLogo software.

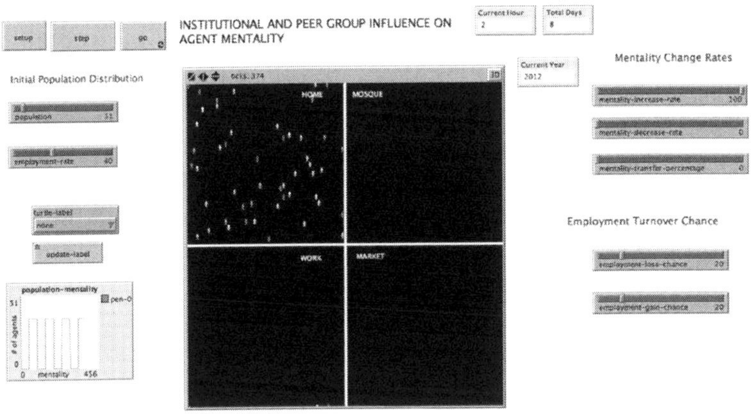

Figure 9.2 NetLogo agent-based model graphical user interface with system in initial state.

In this ABM, the Moore neighborhood is the parameter space of one sight-radius (SR1). The Moore neighborhood is a squared grid space with 8-connected physical points called "pixels," where each neighbor closest to the center pixel is connected to another pixel's edge or corners in the square. The center pixel in the Moore neighborhood is responsible for influencing the neighboring agents in its grid and is linked horizontally, vertically, and diagonally. The concept of sight-radius 1, SR2, and SR3, and so forth is the degree to which the center agent's sight radius can expand its rate of influence. By doing so, I can account for the degree to which an agent travels across the landscape and suggest how an agent's mentality might influence others in terms of their proclivity to achieve an uprising, democracy, or social engineering of any type to reach social constructionism. In this ABM, the ability of an agent to influence its SR is equivalent to having the ability to influence others in a real-life context. In this respect, as the SR increases, it increases the influence parameter of the central agent. Whereas the degree to which agents are influenced is determined by the interaction amongst the agents in the SR bloc. Agents are instructed to complement the highest mentality level in its cell bloc, by trading a lower mentality rate for a higher mentality rate if they interact with an agent with a higher mentality at each time interval. Mentality level is measured as a rate of diffusion among agents in the Moore matrix and it denotes the degree of influence one agent has over another. Mentality level exemplifies the influence of social and religious organizations as well as peers during period preceding the protests in 2011. An agent's mentality heightens by the rate at which it is "influenced" by its peer(s) in the Moore neighborhood. In this model, when mentality level is higher than 50% in one agent, the lesser of the two is instructed to modify to the higher mentality for the remainder of the model run.

9.3.1 Model Dynamics

I gave each agent a definitive set of rules that would reveal the beliefs of one agent on those of another and test the effects of employment on the agent population, using NetLogo software. Employment serves as an independent variable in this model. To achieve this goal, I allowed the agents to respond to a Boolean statement, which asked each agent "if employed, go to work at 8 am," and "if not employed, go to the mosque or the market at 10 am." At every time-tick interval, each agent within the working population in the model was randomly asked whether they were employed.

The degree to which the Moore parameter has an effect on each agent is geared toward simulating either/or the reaction of youth to other individuals, or the submissive agent to an aggressive agent. For example, the agent is conditionally programmed to assume that a higher mentality than its own is superior and therefore takes on that agent's mentality when the influencer agent's mentality is >0.5. Under these conditions, employment has a neutral effect. I find that regardless of an agent's status as employed or non-employed the processing among agents outside of the employment zone has a greater effect on the outcome of each agent and on the macro outcome of the entire agent population. In this sense, an institutional parameter does not seem to affect the agent population in the same way peer groups seem to.

Throughout the entire run-time of the model, the agents continue to influence others on a day-to-day basis throughout intervals of days and years. Their activity is accounted for in the behavior space of NetLogo, which accounts for the agents' activity each day to the next. At this point there is no regression or control factor built in to lower an agent's mentality level, the agents sustain or escalate their mentality each day.

The landscape distribution of the agent population is based on the NetLogo function, random-float 1. This pseudo-random generator disperses the population based on an initial seed variable, which calculates the distribution of the population. Parameters are in place to direct peers into networks. The random-seed generator of the random-float code functions with two independent control variables: the employment rate and the age distribution in the model. The algorithmic distribution of the agent set is a manifestation of the random number generator that serves two purposes. It determines which is more binding, time spent in certain zones or the proximity to radical influences. The radical influence is a measure of the diffusion of mentality in the Moore neighborhood. The random-float 1 creates unpredictably in the ABM. It takes the sum of the population, determined by the global variable on the model interface, and returns a random real value within a range of 0.0 to 1.0 and based off this number agents are allotted within a population age structure and ascribed as employed or non-employed. The random generator poses complexity yet directionality to the model. Thus, establishing a level of employment as a variable establishes a course of action for each agent and controls the rate at which negative directed influence may occur; likewise, it exposes non-employed agents with a greater degree of access to negative influences.

9.3.2 Results

The data in this section are based on multiple runs in NetLogo that test my initial hypothesis that introducing employment as a social structure makes a difference on the agent's continuous behavior. My data suggests that sight radius has an effect on the collective action of the ABMs' agents.

I tested the mean and the standard deviation using descriptive statistics for the effect of the sight radii (SR1–SR20) on each population set of agents at 51, 102, and 500 agents. The data is based on a set of 30 runs for each correlated variable set. For example, my comprehensive analysis correlates SR1–SR20 with populations 6–500 agents, in which I then narrowed the testing parameter to 20 test sets of 30 runs each for populations of 51, 102, and 500. As a result, I ended up with 20 sets of data for the population at 51 and 20 sets of data for the population at 102, and so forth.

Lustik and Miodownik, in their prior contribution to this work, correlated sight radius with tipping points and found that increasing levels of sight radius leads to a series of regular tipping points for a fixed population size [11]. Thus, they credit only one population size to correlate with a tipping point. In their model, the equilibrium point shifts when higher levels of sight radii and create an imbalance in the total agent population set, thus, their model "tips." Their model contains flaws because they used only one single population subset to test their ABMs' tipping point. They affirmed that at high density (2500 + agents), tipping could be observed at very local interactions at SR2 or SR3. I found that among a crowded population density and frequent interactions (at SR1), as well as low-density populations with more expansive coverage of SR, create more tipping points. In addition, my data suggests that not one single SR can be defined and used for public policy or other matters such as crowd behavior because crowds rarely hold the same number of participants and crowd participants vary in personality and social temperature. The value systems and beliefs of the agents are powerful in dense crowds, and crowds vary from culture to culture. Agent-based model cannot account for convergence of crowd tendencies if only one SR is linked to an optimum outcome.

Convergence models host dynamics that are based on the rational choices of individual agents. Convergence is observed when the system exhibits a macro-change of a network facilitated by the step-by-step interactions of the crowd components, i.e., the agents emerge toward a common goal. It is similar to spreading a virus or a rumor on a network. Convergence theory places emphasis on individual-to-individual exchange, holding that individuals guided by both formal and informal rules can change their position in a group and influence the outcome of the interaction. Convergence describes that the mass collective multiplies when each agent makes the choice to replicate the action of another. It is an example of collective peer-pressure that leads to powerful exchanges and creates a collective identity.

In my model, I correlated the significance of the sight radius with increasing radii to demonstrate that the density of agents with high-mentality levels, have a significant effect on neighboring agents. I started each population group at SR1 and compared it with increasing levels of sight radius up to SR20. By comparison, my hypothesis is that if two standard deviations of SR runs are smaller than the difference between the means of those same runs then a convergent pattern occurs within the population group between those two sight radii. The dense clustering of agents is a contributing component. In real-life scenarios, convergence is exhibited when people who choose to behave in similar ways come together to form groups and crowds. What I found significant is that optimality for reaching equilibria at various sight radii is dependent on the number of agents in the population for each run. For instance, at population density 51, 102, and 500 the maximum SR before tips were observed corresponded respectively to SR11, SR11, and SR4. I measure SR1 as the basis point against increasing levels of SR. In a population of 51 agents with a SR12, equilibrium broke and the standard deviation was noted to be 0.19213, which is less than the difference of the two means of two runs (SR1 and SR12) = 0.22014. When I measured the density of 102 agents, their propensity to converge occurred at SR12 with a standard deviation of 0.19213 and the difference between their mean at 0.22014.

The density of the agent set is a contributing factor to achieving meaning behind the statistical scores of each run and the standard deviation between each interaction. Using SR11 and a population set of 500 nothing of significance is observed. Rather, at SR5 and the same population at 500 agents, I achieved a difference of the mean at 0.16729 and the standard deviation at 0.16004. What is interesting to note in terms of the application of ABM models for addressing tipping points for public policy is that not one ABM experiment is significant enough to measure, define, and mandate a direct tip for social scenarios, thus, there is fault in earlier work [12]. We need to build models that expand features of prior work yet integrate new research. We should consider that the density of crowds matter, and SR is significant in all crowds and the occurrence of regular tips will occur as sight radius expands.

Convergent non-emergent patterns should be the focal point for explanations of crowd behavior during the Arab Spring. The convergence of the free riders among the Arab youth during the revolutions created a dynamic of the clustering of ideas and people who were swayed by others in the environment. This pattern is identified in my ABM as agents confirm to other extreme agents in clustering patterns as SR increases. As the distance between agents diminishes, ideas and the social temperature of the crowd diffuse easily between the individual agents. We must pinpoint these fractural, contextual drivers that weaken the social edifice in society. We have seen that parameter space influences agents and when clustering occurs, systems fracture. The proximity to neighboring agents in combination with the size of the population affects the point at which a stable equilibrium breaks. The density and frequency of the interaction in the run, either in real-time terms or in computational models, matter more than highlighting a particular direct tipping point.

My model is not a complete picture but suggests an important aspect of collective action among the unemployed during the Arab Spring. Due to the influencing factor, it suggests that agents were more likely to be exposed to radical ideas outside of the workplace environment. The workplace reduced but did not eliminate radicalization. There are other significant triggers that cannot be observed like the feedback loops and phase transitions that occurred over the course of the protests and even further back in both of the regimes history.

9.4 Referencing Collective Action Theory and Convergent Theories in Agent-Based Modeling

Early theorists on collective behavior saw uprisings as isolated events; they claimed patterns about human nature and collective understanding that suggested groups collectively form when there are provocations from the environment and public sentiment among the crowd. Agent-based models enable us to test this proposition. Due to their ability to process thousands of megabytes of data over an infinite time frame agent interaction in ABM models offers a way to measure transition points from stability to sudden dramatic changes in model landscapes. It can also offer a way to understand the emergent properties of crowd dynamics, and detect the important transitions that change dynamic systems, such as riots. Agent-based model can also assist in formalizing crowd dynamics and answer questions that are not well addressed by large-scale data collection, or by small-scale case study–based analysis.

Gustave LeBon (1895), Gabriel Tarde (1898), Robert E. Park (1921 and 1930), and Herbert Blumer (1946), the first transformation theorists of crowd dynamics, each viewed riots as umbrella systems that should be analyzed as macro phenomena [1,11,14,15,19]. Rather than focusing on the roots of the riot or the individual-to-individual dynamics among crowd participants, their theories laid the groundwork for studying the general theory of collective behavior.[c] However, each of these authors fell short emphasizing the group dynamics and the environmental circumstances that today, are considered contextual drivers of social protests. LeBon, Park, and Blumer focused on the transformation of individuals within a crowd to group consciousness and analyzed crowds on the basis of their social organizational capacity to plan, organize, and coordinate, rather than on the underlying psychological mechanisms of social psychology amongst the interacting agents in the crowd [13: 416]. In understanding the crowd as a dynamic system, sociologists need to account for the context within which a dynamic system tips. To appreciate these dynamic system tips in terms of agent-based modeling we can see that introducing contextual context into a crowd dynamic (such as the dynamics in Egypt and Tunisia prior to the protests) can help shape a hypothesis in computational crowd modeling and forecast tipping points or phase transitions and can help refocus the study of collective action.

Gustave LeBon, Robert Park, and Herbert Blumer's analyses alone cannot help resolve the shortcomings in contemporary crowd understanding [1,11,15]. Their work was criticized for a lack of microanalyses in regards to the complexity of interaction between individuals within a group, in spite of it being praised for their ability to conceptualize the macro picture of collective behavior [13]. Agent-based model allows us to test their earlier collective psychological research empirically and provides a platform to elaborate on their work to analyze individual- to individual agent interaction, which can be used to predict trends toward uprisings and collective action. For this reason, ABM can be used to distinguish

---

[c]Blumer's version of collective action remains "widely circulated in sociological textbooks, military and police crowd control manuals, and popular thought" [13: 408].

structural conditions from triggering events and look at outbreaks of instability among crowds. The goal is to gain an increased ability to predict societal trend reversals that lead from crowd instability to crisis and back to stability.

Identifying collective action in an ABM becomes significant when individuals who initially behave according to informal rules of interaction begin identifying with a large body of agents in a model designed to internalize a group identity and formalize a group ideal. Thus, collective action reflected in an ABM can be observed as a major driver of epistemological social change. The collective behavior in an ABM can also help highlight social phenomena or periods of convergence in computational modeling. Agent interaction can display a convergent process in which shared values, cultural similarities or public sentiment can generate a group identity and achieve a collective goal and distinguish that dynamic from emergence in which the collective tendencies of agents come together to form a collective outcome.

9.4.1 Referencing Further Factors for Future Modeling

The Arab Spring has enabled us to look at changes in systems in terms of collective action and the convergence of group ideals that led to the uprisings. Agent-based models can help analyze how the actions of agent actors impel change in the system, how the model sphere is shaped, and which agents are manipulated by power relationships, ultimately leading us to better understand the process through which collective action occurs and to better understand swings in collective social behavior. When demonstrated in an ABM this can help us shape robust domestic public policy and international policy to strengthen relationships that might help achieve democratic outcomes. Models can help capture and formalize new ideas and new trends that are taking shape in society. They can assist policy makers in understanding the relationships through which people mobilize. These models can serve as learning tools to achieve common knowledge about critical global change processes. Using a model allows us to better understand the dynamics of collective social action to predict trends in crowd behavior. It also assists policy makers better design interventions to solve societal problems that might lead to a crisis or an instability event.

Although many dynamics can be depicted in ABMs there are many that cannot. For example, capturing complexity and phase transitions might be difficult to achieve if the model is programmed to direct actors toward convergent processes.

Analyzing the difference between contextual and direct tips is hard to identify in model environments. To remedy this deficiency, Lustick and Miodownik identify three classifications of tips to suggest a means to measure the change between a micro and a macro tip. A micro tip is considered to change the state of an individual agent, a meso tip is the formulation of self-sustaining clusters, and a macro tip is total convergence of all participants [12]. Their work might be useful to state the degree of influence a contextual trigger might have on a system. Distinguishing between the three types of tips can assist modelers in finding policy geared toward preventing micro and meso tips, and to determine how these tips affect the system at the macro level where they may cause systems to collapse.

Another area that can be explored is how phase transitions within social systems produce tips or trigger instability. Phase transitions are a transition from order to disorder that alters the outcome of a situation [9: 20]. In physics, phase transitions occur when material passes through temperature or atmospheric pressure changes, i.e., ice becomes water as temperature increases and hydrogen gas turns to liquid nitrogen. In non-physical systems, crowd behavior theorists can claim that riot behavior is analogous to a phase transition as the social temperature of the crowd changes from calm to turbulent. Social temperatures increase as peer stimulation inside a group increases, and it can be measured by the degree to which individuals react to information that is supplied by the other crowd participants. Complexity theorists studying the dynamics of class change could argue that phase transitions occur as a result of contextual triggering points in the social environment.

9.5 Conclusion

As discussed in this chapter, the Arab Spring provides a case study for examining links between tipping points and contextual drivers that trigger collective action. The youth bulge is the tipping point that led masses of unemployed youth to protest in January 2011. The youth were frustrated by the fact that they were never the focal point of national development policy. In both Tunisia and Egypt, the middle class made up the bulk of the protesters.

In this agent-based model, I examined the role of collective behavior among the unemployed youth by measuring the degree to which peer group influence affected the actions of the agents in the model. I used peer group influence as a measure to explain why the unemployed and especially the unemployed youth are often susceptible to identifying with radical religious movements that give them a status they cannot achieve in their family or greater society.

I measure tipping points in my model as a response to the pressures of the agent actors influencing others to conform, or converge to the group norm. The network clustering can be correlated with degrees of radicalization within the sight-radius of the agent set and my model demonstrates that convergent patterns that drive network clustering can explain a degree of radicalization among an agent subset. This radicalization can then help us understand the structure and motivation of the social networks that impelled the Arab revolution in 2011.

References

1. Blumer, H. (1946). Collective Behavior, in *Principles of Sociology* (ed., Lee, A. M.) New York: Barnes & Noble, pp. 67–121.
2. Bos, N., Greenberg, A., Kopecky, J., Ihde, A., Simpkins, S. (2011). Sociocultural Modeling in the Social Identity Look-Ahead Simulation and the Green Country Model. *John Hopkins Applied Technical Digest,* **30**(1), 13–21.
3. Dalacoura, K. (2012). The 2011 Uprisings in the Arab Middle East: Political Change and Geopolitical Implications. *International Affairs,* **88**(1), 63–79.
4. Gillespie, M. (2013). BBC Arabic, Social Media, and Citizen Production: An Experiment in Digital Democracy before the Arab Spring. *Theory, Culture & Society,* **30**(4), forthcoming July/Summer 2014.
5. Granovetter, M. (1978). Threshold Models of Collective Behavior. *American Journal of Sociology,* **83**, 1420–1443.

6. Gray, J. (2012). Doing Business Stumbling Blocks: The Arab Spring Has Been Touch on Egypt When It Comes to Reforming Their Business Environment. *News in Focus, Business Today.*
7. ITUC Survey. (2012). *Tunisia*. Accessed: http://survery.ituccsi.org/tunisia.html?lang=en.
8. LaGraffe, D. (2012). The Youth Bulge in Egypt: An Intersection of Demographics, Security, and the Arab Spring. *Journal of Strategic Studies,* **5**(2), 65–80.
9. Lamberson, P. J., and Page, S. E. (2012), Tipping Points, *Santa Fe Institute Working Paper*, pp. 1–30.
10. Lang, J., and De Sterck, H. (2012). The Arab Spring: A Simple Compartmental Model for the Dynamics of a Revolution. Unpublished, *Working Paper*, Department of Mathematics, University of Waterloo.
11. LeBon, G. (1895). *Psychologie des Foules* (Paris: Alcan). Translated (1960) *The Crowd* (New York: Viking Press).
12. Lustick, I., and Miodownik, D. (2004). Everyone I know Is Doing It: Tipping, Political Cascades and Individual Zones of Knowledge, *Proc. A. Meeting APSA,* September 1–5.
13. McPhail, C. (1989). Blumer's Theory of Collective Behavior: The Development of a Non-Symbolic Interaction Explanation. *The Sociol Quart,* **30**(3), 401–423.
14. Park, R., Burgess, E. (1921). *Introduction to the Science of Sociology* (Chicago: University of Chicago Press).
15. Park, R. (1930). Collective Behavior, in *Encyclopedia of Social Sciences,* vol. 3 (ed. Seligman, E.) New York: Macmillan.
16. Pelechanco, N., Allbeck, J., and Badler, N. (2008). *Virtual Crowds: Methods, Simulations, and Control Synthesis Lectures on Computer Graphics and Animation* (Morgan & Claypool Publishers and Atypon Systems Inc.).
17. Spitzberg, B., Gupta, D., Tsou, M. H., An, L., and Gawron, J. (2012). The Map Is Not Which Territory? Speculating on the Geo-Spatial Diffusion of Ideas in the Arab Spring of 2011. *Working Paper, NSF,* pp. 1–31.
18. Tadros, M. (2012). Introduction: The Pulse of the Arab Revolt, *IDS Bulletin,* **43**(1), 1–15.
19. Tarde, G. (1969). The Public and the Crowd, in *On Communication and Social Influence* (ed. Clark, T. N.), Chicago and London: University of Chicago Press, pp. 277–296.

20. Turner, R. (1964). Collective Behavior. In *Handbook of Modern Sociology* (ed. Faris, R. E. I), Chicago: Rand-McNally, 382–425.
21. Wardany, Y. (2012). The Mubarak Regime's Failed Youth Policies and the January Uprising. *IDS Bulletin*, **43**(1), 37–46.
22. Wolfram MathWorld. Accessed February 8, 20132. Retrieved from http://mathworld.wolfram.com/MooreNeighborhood.html.

Chapter 10

Complexity of Structure in Electronic Medical Record Systems

Russell S. Gonnering,[a] **Gregory S. H. Ogawa,**[b] **Sven-Erik Bursell,**[c] **and Mark B. Horton**[d]

[a]*Department of Ophthalmology, The Medical College of Wisconsin, 1780 San Fernando Drive, Elm Grove, WI 53122, USA*
[b]*Eye Associates of New Mexico, 8801 Horizon Blvd., #360, Albuquerque, NM 87113, USA*
[c]*Telehealth Research Institute, MEB, The John A. Burns School of Medicine, University of Hawaii, 651 Ilao Street, Honolulu, HI 96813, USA*
[d]*IHS-JVN Teleophthalmology Program, 4212 North 16th Street, Phoenix, AZ 85016, USA*

rsgonnering1@me.com

Through a combination of governmental and private incentives, a transformation of medical records into an electronic medical record (EMR) format is spreading rapidly. Proponents of transformation to the EMR promised a number of improvements in healthcare, but disturbing signs of possible adverse effects have surfaced. In particular, we are concerned about the inability of the current records to adequately communicate the complex, information-in-context contained in many medical records as well as the emergence of diagnosis that results from time-dependent analysis and review of a problem. We investigated the complexity of structure in three EMR as an initial step, and contrasted these findings with those

Complexity and the Human Experience: Modeling Complexity in the Humanities and Social Sciences
Edited by Paul A. Youngman and Mirsad Hadzikadic

ISBN 978-981-4463-26-3 (Hardcover), 978-981-4463-27-0 (eBook)
www.panstanford.com

in transcribed paper records, dictated reports, and articles in biomedical journals. We found a marked difference in function words, suggesting a limitation on the transfer of information-in-context in the EMR. When the wide choice of narrative and its extensive vocabulary is exchanged for limited choices in a list, the length of the message must be extended to transmit the same amount of information. This limitation on contextual information in turn has subtle, yet profoundly significant, implications for the behavior of physicians and other healthcare professionals.

10.1 Introduction

The application of EMR to healthcare has long been the goal of governments in the United States and abroad. The 2009 American Recovery and Reinvestment Act set the goal of utilization of an EMR for every citizen by 2014.[1] The Congressional Budget Office estimated the health IT provisions of this bill would amount to 18 billion USD added to the deficit over 11 years.[2] Proponents of the EMR point to benefits that will result from this adoption[3,4]:

- complete and accurate information
- better access to information
- patient empowerment
- better decision making
- overall cost savings

Greenhalgh and co-authors question, however, whether these benefits will actually accrue. Moreover, they are concerned about the potential unconscious biases, tensions, and paradox in published studies on the subject.[5] Of particular interest is the tension between "knowledge as transferable fact" and "knowledge as information-in-context." A positivist approach to EMR assumes the former, and that the decision support offered by EMR will eventually improve quality, even though the evidence has not shown it yet. The alternative view is that major advances in quality are unlikely to occur as healthcare is filled with contextual information rife with exceptions to general rules. The view that healthcare is filled with exceptions and contextual information is bolstered by the numerous articles that describe healthcare delivery as a complex adaptive system (CAS).[6–8] Such research begs the question: Do the records of healthcare share this understanding of the primary activity of healthcare delivery being a CAS? We think not.

Complex, contextual information is most easily transmitted through narrative.[9] It is well recognized by authors such as Greenhalgh and Hurwitz that the study of narrative in medicine offers an added dimension of understanding the context of disease.[10,11] Thus, the most accurate and fullest transfer of contextual information should be an important aspect of any EMR. Unfortunately, this is precisely the area that is lost in many iterations of EMR.[12] Template-driven drop-down lists take the place of dictated descriptions. We question if the limitations in the structural complexity of EMR is a restrictive factor in the communication of the complexity of content.

The medical record is a means for asynchronously transmitting the context-rich information contributed by the patient in the history, as well as the equally context-rich information produced as a result of the physical examination on the part of the various providers. Additional contextual information elicited by ancillary studies is subsequently added. Each step of the process is highly influenced by the starting point and non-linear relationships with multiple factors. At each step, the availability of information as well as its quality will influence the subsequent course of the process. Treatment "emerges" from an evaluation process steeped in oral narrative.

What is altered by the change from an oral narrative to a written account? Most physicians would think very little. However, the effect of the medium on the message has been the focus of study by the members of what has become to be known as the Toronto School of Communication.[13] Perhaps the best known of this school of thought, Marshall McLuhan, coined the famous phrase "the medium is the message" to illustrate the influence the communication technology had on the content of the communication, as well as the thought processes of the originator.[14]

In *The Alphabet Effect: A Media Ecology Understanding of the Making of Western Civilization*, Robert Logan, another member of the Toronto School, followed the development of communication technology.[15] He traced the beginning of "written" communication from marks on tally sticks in hunter-gatherer societies of 15,000–

8,000 BC. Each transition to different media corresponded to break-boundaries between different economic and social patterns, and the need to communicate ever increasing contextually complex information (Table 10.1):

Table 10.1 Medium transition based upon complexity of message

Time frame	Medium	Message	Cultural transition
15,000–8,000 BC	Tally stick	Count	Hunter-gatherer
8,000–3,400 BC	3-Dimensional token	Quantity and quality	Agriculture
3,400–3,250 BC	Complex token	Metaphor	Early urbanization
3,250–3,150 BC	Token-impressed envelope	2-Dimensional representation	Increased trade
3,150–3,100 BC	Impressed tablet	Abstract thinking	Metallurgy
3,100–3,900 BC	Incised tablet	Science, philosophy	Logographic writing
1,900 BC	Proto-Sinaitic script	Theology, individualism	Phonetic writing

A glance at this table bears striking conclusions: We have taken a monumental step backwards 5,000 years in the structural complexity of medical records In effect, the practitioner is moving backwards on the Logan timeline to the use of "digital tokens."

Many of the commercially available EMRs use some sort of menu-driven aid to produce notes that formerly were either hand-written or dictated. These sometimes combine the ability to use drop-down lists as well as free-form text. Some also add drawing capabilities. In theory, this should allow the author to produce a note of equal, if not superior, depth and clarity of information than that produced by prior means. However, theory does not seem to meet practice.

At a recent medical meeting on the impact of EMRs, one of the attendees related that a colleague's note, which used to be a model of clarity, now reads like that of a Third-Year medical student. With the time constraints of modern medicine, and the decrease in clinical productivity attendant with transition to an EMR, the temptation is to use the drop-down menus to produce an "adequate" note. Perhaps of even more significance is the use of "cut and paste" to copy prior notes in part or in toto. The result is a record of dubious validity and utility.[16]

Nuance and context may also take a backseat due to the choice of which digital tokens are made available by the programmers of the EMR. For instance, my own practice involves treatment of three different movement disorders: Essential Blepharospasm, Hemifacial Spasm, and Meige Syndrome. These choices are available on the list of discharge diagnoses, but my only choice for chief complaint is "nerve problem (facial nerve)." This complaint covers a vast array of problems completely unrelated to the conditions for which I am seeing the patient, yet it is the only choice available to me. Future researchers looking for patients with the complaint of facial spasm will be stymied, as that complaint is not even listed. They will need to understand that in this particular EMR, if they are looking for records of patients with a complaint of facial spasm, they will need to search for "nerve problem (facial nerve)." The structure of the EMR has forced the creation of a "digital idiom" or dialect peculiar to the programmers at this institution. How will such idioms be understood 5 years from now? Twenty-five years from now? In addition, the implementation of EMRs can produce multiple unintended consequences.[17] Not the least of these is the potential for lack of clarity and confusion if the medical records are required for legal proceedings.[18]

Of perhaps even more potential significance relating to the type of information captured and methodology of information transfer in EMRs is the impact on safety.[19] The preface of the prepublication copy of the Institute of Medicine report, *Health IT and Patient Safety: Building Safer Systems for Better Care*, notes the following:

> We found that specific types of health IT can improve patient safety under the right conditions, but those conditions cannot be replicated easily and require continual effort to achieve. We tried to balance the findings in the literature with anecdotes from the field but came to the realization that the information needed for an objective analysis and assessment of the safety of health IT and its use was not available. This realization was eye-opening

> and drove the committee to consider ways to make information about the magnitude of the harm discoverable.[20]

Objective standards for analysis of complexity of information conveyed in text are difficult to find. This perhaps is not surprising, as the faculty of the Santa Fe Institute has not been able to agree on what is meant by "complexity" in the first place.[21] While Gell-Mann and Lloyd have suggested concepts such as Effective Complexity (EC) and Algorithmic Information Content (AIC) as a means of measuring complexity of character strings and language,[22,23] the utility of these concepts seems minimal in this context. One can imagine a text in French to have the same EC and/or AIC as a text in English, but the former would be completely unintelligible unless the reader knew French. Existing research tools for textual analysis may be a starting point.

10.2 Methods

We collected digital records from three EMR systems and stripped them of protected health information. These included a commercially available system (A-Electronic, 15 records), a proprietary system (B, 71 records) and a system written by one of the authors (C, 268 records). These were compared to transcribed paper records from the same institution using A (A-Paper, 15 records), dictated consultation reports (Dictated, 12 records) and 30 articles in the Biomedical collection (Biomed, 30 records) of the American National Corpus.[24]

We used two software programs to analyze the records:

Linguistic Inquiry and Word Count (LIWC) is a research tool developed primarily to use text analysis as a means of counting frequency of various words and parts of speech and possible attributes of the author based on that frequency.[25] Built upon study of over 168 million words (composed by 24,000 writers in six classes of text, including scientific writing, in 72 studies) the program provides an analysis of target text according to 80 different parameters.[26]

Word Counter is a program used to build word-frequency tables for additional study.[27]

We then calculated the Shannon entropy and LMC statistical complexity for each of the record systems.

10.3 Results

In Table 10.2, we provide the number of records, the word count, the number of unique words (words used only once) and the percentage of unique words.

Table 10.2 Number of records, total word count and unique words

Record	Number of records	Total word count	Unique words	Percent unique words (%)
A—Electronic	15	12,106	1,256	9.6
B	71	96,371	1,981	2.0
C	268	62,029	3,491	5.6
A—Paper	15	3,987	730	18.3
Dictated	12	9,318	2,111	22.7
Biomed	30	114,250	9,495	8.3

The word frequencies followed a long-tail distribution in all cases (Figs. 10.1 and 10.2). This is not surprising, as Li has demonstrated that an inverse power law is operative both in natural languages as well as in randomly generated texts.[28] What is unusual, however, is the makeup of the words in each of the examples.

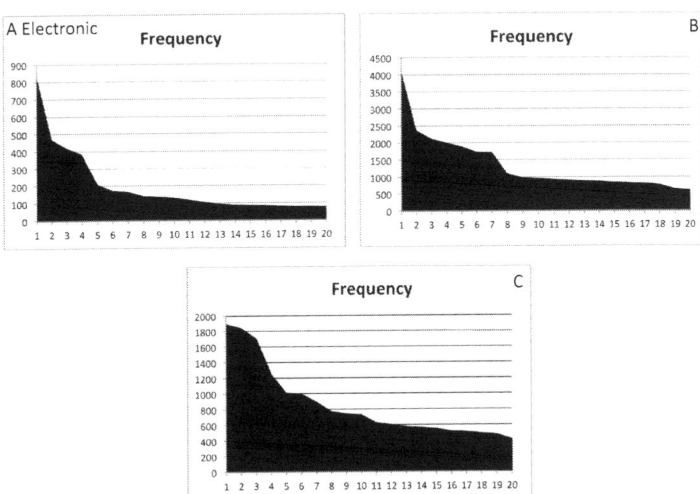

Figure 10.1 Graph of frequency of first 20 words in EMRs.

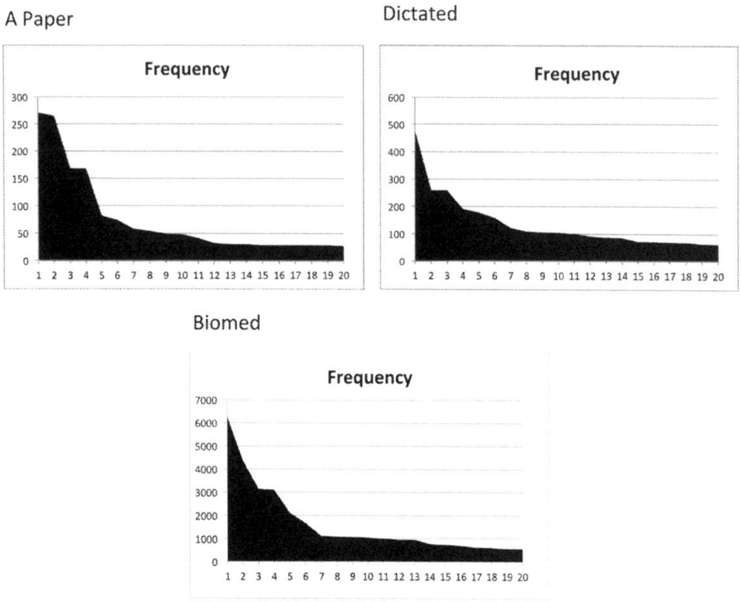

Figure 10.2 Graph of frequency of first 20 words in non-EMR records.

Chung and Pennebaker reviewed the literature on what are termed "function words," such as pronouns, prepositions, articles, conjunctions, and auxiliary verbs.[29] These words account for less than 0.04% of the average vocabulary, yet account for over half of the words used in normal speech and are the hardest words in a foreign language to master. They are used almost unconsciously, yet are extremely important in giving context to the "content words" of the message. Figure 10.3 shows the relative frequency of function words in the records studied.

The situation is somewhat reversed when percentage of numerals in the record is analyzed, as shown in Fig. 10.4.

The records in A—Electronic, A—Paper and B had a very low percentage of function words, while the Dictated and Biomed records had a relatively high percentage. The C records were intermediate, perhaps due to the free text option that was readily available in this program. A–Paper, A–Electronic, and B conversely had a significantly higher frequency of numerals, whereas C, Dictated, and Biomed each had similar and significantly lower, frequencies.

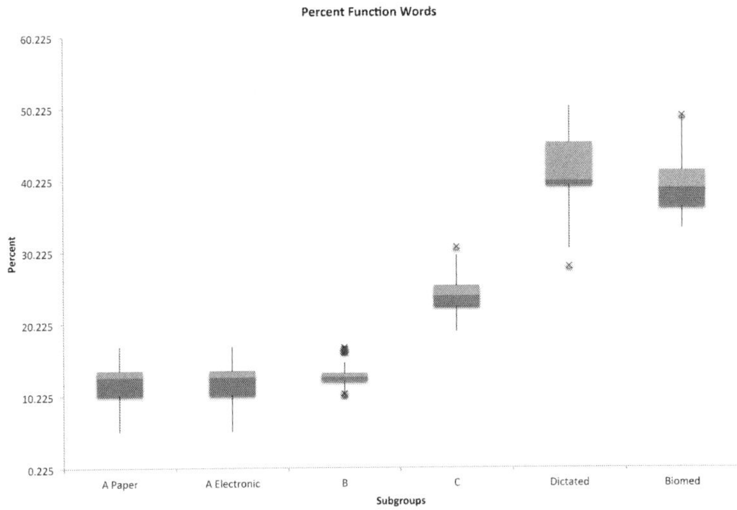

Figure 10.3 Percentage of frequency of "function words" in records studied.

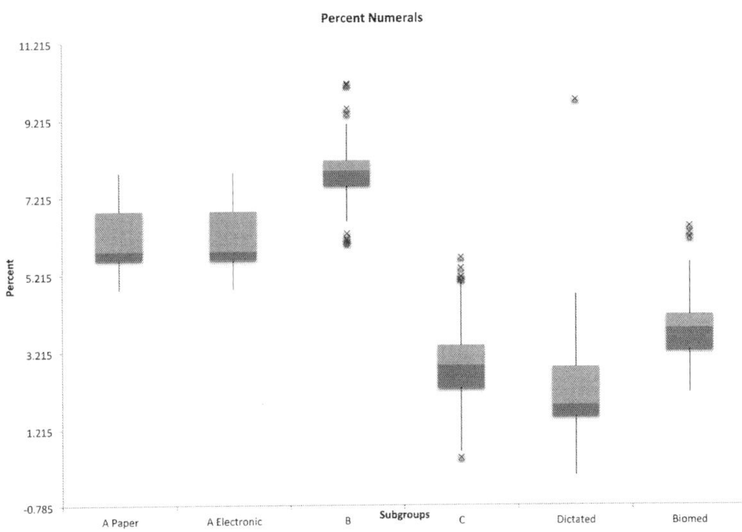

Figure 10.4 Percentage of frequency of numerals in records studied.

The top 10 words in each record are demonstrated in Table 10.3.

Table 10.3 Top 10 words in each record

Biomed	Dictated	C	A—Paper	A—lectronic	B
The	The	The	OD	=	7
Of	And	Normal	OS	OD	20
In	Of	No	+	OS	10
And	To	And	Normal	Normal	3
To	In	With	20	Mod	11
A	A	Of	2	+	6
That	Is	Mm	1	1	5
Is	With	Right	×	Patient	12
With	Was	Negative	SLE	Negative	Medical center
For	Right	Left	Reviewed	Date	2

In an attempt to quantify the information contained in a system, Shannon and Weaver described what has come to be known as Shannon entropy.[30]

$$H = -K \sum_{i=1}^{N} p_i \log p_i,$$

where H = Shannon entropy measured in Nats, K = 1, N = number of items in the system, and p = probability of item occurring in the system.

When Shannon entropy is "normalized" by dividing the above equation by the log of the number of words in the subject text, a value between 0 and 1 is obtained. A value of 0 indicates a totally predictable outcome, as in the string, "aaaaaaaaaaaaa." A value of 1, on the other hand, indicates a totally random outcome, as in the string "axvwnpcqidsze." Maximum structural complexity of the message is in the midrange between these two values.

Lopez-Ruiz, Mancini, and Calbet introduced what they called a measure of "statistical complexity" by multiplying the Shannon entropy equation by a factor they termed "disequilibrium," or the distance of the target collection from an equi-probable distribution.[31] Combining their disequilibrium factor with Shannon entropy gives the following value for statistical complexity:

$$C = HD$$
$$= -\left(K\sum_{i=1}^{N} p_i \log p_i\right)\left(\sum_{i=1}^{N}(p_i - 1/N)^2\right),$$

where C = LMC statistical complexity, H = Shannon entropy, D = disequilibrium. K = constant (1 when measurements are in Nats) N = number of items in the system, and p = probability of item occurring in the system.

Statistical complexity is not univocal, as many distributions can have a different value of C for the same value of H. As N goes towards infinity, the maximum complexity for a given normalized H becomes[32]:

$$\overline{C}_{\max} = \overline{H} \times (1 - \overline{H})^2$$

Table 10.4 gives the results for normalized Shannon entropy, LMC statistical complexity, and percentage of maximum LMC statistical complexity for each of the record systems investigated.

Table 10.4 Normalized Shannon entropy, LMC statistical complexity and percent maximum LMC statistical complexity for the record systems investigated

Record	Normalized shannon entropy	LMC statistical complexity	Percent maximum LMC
A—Paper	0.685	0.040	42.22
Dictated	0.703	0.020	67.61
A—Electronic	0.536	0.031	73.08
Biomed	0.607	0.024	74.44
B	0.544	0.018	83.79
C	0.566	0.017	84.04

10.4 Discussion

We found a marked stratification of records based upon the percentage of function words. The highest percentage, as probably could be expected, occurred in the dictated consultation letters and biomedical journals. The lowest percentages were in two of the EMRs as well as the transcribed paper records. As the paper records made extensive use of abbreviations and fragmented sentence structure, their skewing towards the EMR group can be explained. One EMR exhibited an intermediate structure, most likely explained by the generous use of free text fields.

In Logan's framework, function words, so common in spoken language, would only reach their full expression in written communications by the use of a phonetic alphabet. The earlier token-based and logographic written communication media could express content accurately, but the *context* was limited. Finally, contextual information could be shared across distances in time and space.

The marked decrease in function words is consistent with, and helps to explain, the concern regarding the loss of the sense-making narrative in the EMR. In general, when the wide choice of narrative and its extensive vocabulary is exchanged for limited choices in a list, the length of the message must be extended to transmit the same amount of information. In 1928, Ralph Hartley described his equation on transmission of information based upon the string length and number of possible symbols used in the transmission system[33]:

$$\overline{H} = \log S^n,$$

where H = information, S = possible symbols from which to pick, n = symbols in the information string.

When a choice list or menu is used to construct a quasi-narrative, the amount of information in a given string necessarily decreases. In his book on information, James Gleick uses the example of the talking drums of Africa to demonstrate this property.[34] Because the drums have only a small number of symbols compared to spoken language, their message must be almost 8 times as long in order to transmit the same information. Likewise, when limited lists are used to construct a medical record, the length of the record must necessarily be longer to convey the same amount of information. However, they rarely are. Therefore, the information content of the record must be less, or potentially less, than the alternative.

Anecdotal evidence from around the country abounds that bears this out. Physicians complain that all the medical records for a given diagnosis look alike. They are flooded by data, but have very little useful information for some diagnoses. Much of the "content" of the medical record is ignored as one searches for the few sentences that will "make sense" of the encounter.

Yet in some instances, the EMR is profoundly useful, as it allows tracking and clear demonstration of vital content when that content is needed to deliver quality care. Why is that? The answer can perhaps be found by reaching an understanding of the difference between "complicated" and "complex" information.[35] Complicated information responds to a reductionist approach. It is the domain of experts, and knowledge in that domain consists primarily of transferable facts. Complex information, on the other hand, demands context in order to be useful. A reductionist approach is not usually helpful as cause and effect are only seen in retrospect. The tension between these two types of information, their relationship to two types of disease processes and appreciation by two types of evaluators no doubt accounts for the considerable tension and paradox involving the present EMR use.

The Shannon Entropies of the record systems showed that the three EMR systems tended towards the midrange, or region of more potential structural complexity. The LMC statistical complexity studies are more difficult to interpret. Two of the EMRs scored closest to the maximum LMC statistical complexity for their Shannon entropy. The third was only slightly behind. The very low scoring of the transcribed paper record is perhaps due to its fragmented nature. The intermediate scoring of the dictated reports and biomedical journals, though, is difficult to explain. It may be that there is no statistically significant difference between the records when the transcribed paper records are eliminated. A repeat of the study measuring the Shannon entropy and LMC statistical complexity of *each record*, rather than aggregated into types, may show a difference. Likewise, utilization of a different statistical complexity calculation to obviate some potential disadvantages of the LMC may give more useful information.[36]

Further investigation is needed to evaluate what is now only anecdotal information regarding the alteration of physician behavior based upon the adoption of the EMR. Building on the ideas of McLuhan, Neil Postman investigated how the medium of information transfer affects the agents of transmission as well as society as a whole.[37] To a large degree, *the work-product of the physician has become the production of the EMR*. While there is little evidence this is currently adversely affecting quality of care, systematic study is needed. At the very least, a tolerance for a growing imprecision of the recording of information is spreading. Such tolerance will have profound effects upon future clinical research based upon the recorded information in the future. One of the major benefits of the EMR, the easing of clinical investigation, may be compromised by a cheapening of the very information it easily stores. Thus, in the future, we may be able to easily retrieve information of questionable value.

There are hopeful signs on the horizon, however. The application of Natural Language Processing to narrative in the medical record may provide a viable alternative to the current EMR programs, and avoid some of the problems we present.[38] Data mining using computational linguistic tools can populate data fields in an EMR while still retaining the richness of information-in-context in a clinical narrative. Such tools are commercially available, even if not yet widely disseminated.[39]

This investigation has focused on the structure of the EMR. While other authors have used structural complexity[36] and function words[40] to analyze works of literature, to our knowledge this is the first study to apply those tools to the medical record. Additional work is ongoing on the analysis of the contextual complexity of the medical record. However, as a study of Table 10.1 shows, increasing contextual complexity requires increasing structural complexity to support it. The medical record of the future must contribute that structural complexity if it is to be a useful tool.

References

1. American Recovery and Reinvestment Act of 2009, Subtitle A—Promotion of Health Information Technology. http://frwebgate.access.gpo.gov/cgi-bin/getdoc.cgi?dbname=111_cong_bills&docid=f:h1enr.pdf. Accessed 1/16/2012.
2. Congressional Budget Office. http://cbo.gov/ftpdocs/99xx/doc9968/hr1.pdf. Accessed 1/16/2011.

3. Office of the National Coordinator for Health Information Technology. http://healthit.hhs.gov/portal/server.pt?open=512&objID=2996&mode=2. Accessed 1/16/2012.

4. Blumenthal D., Tavenner M. (2010). The "meaningful use" regulation for electronic health records, *NEJM*, **363**, 501–504.

5. Greenhalgh T., Potts H., Wong G., Bark P., Swinglehurst D. (2009). Tensions and paradoxes in electronic patient record research: a systematic literature review using the metanarrative method, *Mil Quart*, **87**, 729–788.

6. Plsek P. (2001). Appendix B-Redesigning health care with insights from the science of Complex Adaptive Systems, in *Crossing the Quality Chasm: A New Health System for the 21st Century*, Institute of Medicine, National Academy Press, Washington, D.C., pp. 309–322.

7. Plsek P. E., Greenhalgh T. (2001). Complexity science: the challenge of complexity in health care, *BMJ*, **323**, 625–628.

8. Wilson T., Holt T., Greenhalgh T. (2001). Complexity science: complexity and clinical care, *BMJ*, **323**, 685–688.

9. Browning L., Boudès T. (2005). The use of narrative to understand and respond to complexity: a comparative analysis of the Cynefin and Weickian Models, *E: CO*, **7**, 35–42.

10. Greenhalgh T., Hurwitz B. (1999). Why study narrative? *BM*, **318**, 48–50.

11. Greenhalgh T. (1999). Narrative based medicine: narrative based medicine in an evidence based world, *BMJ*, **318**, 323.

12. Mace S. (2012). Are EMR's killing the clinical narrative? *Health Lead Med*, March 20. 2012. http://www.healthleadersmedia.com/page-1/TEC-277956/Are-EMRs-Killing-the-Clinical-Narrative. Accessed July 2, 2012.

13. Watson R., Blondheim M. (eds) (2007). *The Toronto School of Communication Theory: Interpretations, Extensions, Applications*. Hebrew University Magnes Press, Jerusalem.

14. McLuhan M. (1994). *Understanding Media: The Extensions of Man*, MIT Press, Cambridge.

15. Logan R. (2004). *The Alphabet Effect: A Media Ecology Understanding of the Making of Western Civilization*, Hampton Press, Cresskill, NJ.

16. Hirschtick R. E. (2012). John Lennon's Elbow, *JAMA*, **308**, 463–464.

17. Jones S. S., Koppel R., Ridgely M. S., Palen T. E., Wu S., Harrison M. I. (2011). *Guide to Reducing Unintended Consequences of Electronic*

Health Records, Prepared by RAND Corporation under Contract No. HHSA290200600017I, Task Order #5. Agency for Healthcare Research and Quality (AHRQ). Rockville, MD.

18. Dimick C. (2010). HER's prove a difficult witness in court, *J AHIMA,* Sept 24 2010 online edition, http://journal.ahima.org/2010/09/24/ehrs-difficult-witness-in-court/. Accessed 1/16/2012.
19. Terhune C., Epstein K., Arnst C. (2009). The dubious promise of digital medicine, *Bussiness Week*, May 4, 2009. http://www.businessweek.com/magazine/content/09_18/b4129030606214.htm. Accessed 2/24/2012.
20. Warden G. L., Bagian J. P., Baron R., Bates D. W., et al. (2012). *Health IT and Patient Safety: Building Safer Systems for Better Care*, Institute of Medicine, National Academy Press, Washington, D.C. (prepublication copy available at: http://www.ischool.drexel.edu/faculty/ssilverstein/Patient%20Safety%20and%20Health%20IT%20prepub.pdf). Accessed 1/20/2012.
21. Mitchell M. (2009). *Complexity: A Guided Tour*. Oxford University Press, New York, NY.
22. Gell-Mann M. (1995). What is complexity? *Complexity,* **1**, 16–95.
23. Gell-Mann M., Lloyd S. (1996). Information measures, effective complexity and total information, *Complexity,* **2,** 44–52.
24. http://www.americannationalcorpus.org/OANC/. Accessed 1/26/2012.
25. Tausczik Y. R., Pennebaker J. W. (2010). The psychological meaning of words: LIWC and computerized text analysis methods, *J Lang Soc Psych*, **29**, 24–54.
26. http://www.liwc.net/howliwcworks.hp#index1.
27. http://www.supermagnus.com/mac/Word_Counter/.
28. Li W. (1992). Random texts exhibit Zipf's-Law-like word frequency distribution, *IEEE Trans Infor Theory,* **38**, 1842–1845.
29. Chung C., Pennebaker J. (2007). The psychological functions of function words, in *Social Communication,* (ed Fiedler K), Chapter 12, Psychology Press, New York, pp. 343–359.
30. Shannon C. E., Weaver W. (1949). *The Mathematical Theory of Communication,* University of Illinois Press, Urbana, IL.
31. Lopez-Ruiz R., Mancini H. L., Calbet, X. (1995). A statistical measure of complexity, *Phys Let A,* **209**, 321–326.
32. Lopez-Ruiz R. (2005). Shannon information, LMC complexity and Renyi entropies: straightforward approach, *Biophys Chem,* **15**, 215–218.

33. Hartley, R. (1928). Transmission of information, *Bell Sys Tech J*, **7**, 535–563.

34. Gleick, J. (2011). *The Information: A History, A Theory, A Flood*, Vintage Books, New York.

35. Snowden D., Boone M. E. (2007). A leader's framework for decision making, *Har Bus Rev*, **85**, 68–76.

36. Rosso O. A., Craig H., Moscato P. (2009). Shakespeare and other English Renaissance authors as characterized by information theory complexity quantifiers, *Phys A*, **388**, 916–926.

37. Postman N. (1985). *Amusing Ourselves to Death: Public Discourse in the Age of Show Business*, Viking Penguin, New York.

38. Resnik P. (9–13 September 2012). *Language Technology and the Clinical Narrative*, paper presented at the South by Southwest Conference, Austin, TX.

39. http://www.coderyte.com/node/128.

40. Hughes J. M., Foti N. J., Krakauer D. C., Rockmore D. N. (2012). Quantitative patterns of stylistic influence in the evolution of literature, *PNAS*, **109**, 7682–7686.

Chapter 11

Adaptation of Foodways and Indicators of Malnutrition: Philippines as a Case Study

Amanda L. Andrei

Department of Computational Social Science, George Mason University, 4400 University Dr, Fairfax, VA 22030, USA

aandrei@gmu.edu

While often associated with poverty and starvation, malnutrition depends on a variety of social, economic, and health factors and is measured with many different criteria, including energy requirements, food supply, food consumption, anthropometric data, and micronutrient deficiencies. Using the National Capital Region of the Philippines as a case study, this project analyzes the distribution of food consumed across different socioeconomic classes, using four criteria to determine a multi-dimensional composite metric of malnutrition in order to compare malnutrition across different classes. Metrics considered are the ratio of macronutrients, the absolute number of grams of protein, the absolute number of grams of fat, and the absolute number of calories as energy intake. The author discovered that the distribution of types of food consumed per capita follows a power law, which

Complexity and the Human Experience: Modeling Complexity in the Humanities and Social Sciences
Edited by Paul A. Youngman and Mirsad Hadzikadic

ISBN 978-981-4463-26-3 (Hardcover), 978-981-4463-27-0 (eBook)
www.panstanford.com

decreases in accuracy of fit as the socioeconomic status of the class decreases. The indicators of malnutrition indicate that common features of malnutrition are prevalent across all classes, while power law analysis provides useful insights into recommendations for managing food security and stability.

11.1 Introduction

The National Institute of Health defines Malnutrition as as "the condition that occurs when your body does not get enough nutrients" [1] and encompasses not only undernourishment but also over-nourishment, as well as lack of vitamins and minerals in a diet. The Food and Agriculture Organization of the United Nations (FAO) and World Health Organization estimate that roughly 30% of the world population experiences malnutrition [2,3]. Furthermore, undernourishment and micronutrient deficiencies are concentrated in Asia more than any other continent [4]. The social, economic, and political heterogeneity across countries makes it beneficial to profile countries or regions instead of attempting to generalize across vastly different cultures. For an example of such diversity, one metric for under-nourishment, the Global Hunger Index (GHI), deemed that most Southeast Asian countries in 2010 scored a GHI between 10.0 and 19.0 and labeled the conditions of the nations as "Serious." In comparison, countries in South Asia were generally labeled as "Alarming" and countries in East Asia often scored a "Low" GHI or were classified as industrialized nations [5].[a]

I chose the Philippines as a case study for this project for two main reasons. First, the country has a unique historical status within Asia as an early European colony, which has produced a culture and food system unlike other Southeast Asian countries [6]. Another reason was the availability of data, as the Philippines is one of two Asian countries[b] participating in FAO's Country STAT project to reach global coverage on providing reliable and readily available agricultural statistics. Furthermore, the National Capital Region (NCR) of the Philippines is home to over 11.5 million people, roughly 13% of the national population, and is the 11th most populous metropolitan area in the world [7], yet it appears that the caloric

[a]With the exception of North Korea.
[b]The other country is Bhutan.

consumption in urban areas is less than that of rural areas and that the urban poor may suffer from more nutritional deficiencies [8]. In addition to other food security conditions, obesity and protein-energy malnutrition (PEM) are two major concerns and areas for local, regional, and national policymaking in the Philippines [9,10].

Most complexity theory research on food systems deals with food webs and food networks within ecological systems [11]. In such studies, power law analysis is limited and pertains mostly to the diets of animals [12,13]. Of power law findings in social food systems, one study determined that recipes in several cookbooks from different nations and historical periods followed a power law distribution [14]. It appears that the study of food as a social system, not simply as a natural or agricultural system, is a potentially rich area of research to apply complexity theory.

11.2 Method of Analysis

The Bureau of Agricultural Statistics of the Philippines (BAS) in partnership with the FAO compiled the dataset for this study. It was part of a larger source of data about yearly per capita consumption of agricultural commodities across socioeconomic classes, which included information for two sets of years, 1999–2000 and 2008–2009, as well as every region within the country and the Philippines as a whole. This particular dataset included amount and types of agricultural commodities, as well as certain processed foods, consumed by the upper, middle, lower, and extremely lower classes of the NCR in the years 2008–2009. This analysis involved two frameworks: four indicators of nutrition to assess and compare health across socioeconomic classes, and power laws as a class of models to derive implications and recommendations for the social food system as a whole.

The four indicators of nutrition provide a general view of malnourishment, as they focus on the macronutrients of carbohydrates, fats, and proteins and the units of energy as calories.[c] The first indicator, the ratio of carbohydrates to fats to proteins measured in grams consumed per person per year, is calculated across classes and then compared to the standard recommended

[c]Calories in this case refer to the "big calorie" or "dietary calorie," also known as the kCal, which is equivalent to 1000 "small calories" or 4.2 kilojoules.

diet for healthy, active adults, which is assumed to be 60% carbohydrates, 25% fats, and 15% proteins, adding up to a minimum of 2000 calories per day [15]. It is important to note that the standard ratio of macronutrients varies across factors such as gender, age, health (active, convalescent), and in the case of women, whether or not they are pregnant.

The second indicator, total intake of calories per person per year, is calculated by multiplying the kilograms of each food item consumed by the calories per kilogram in that food item, and summing across all foods. I encountered some ambiguity in determining the appropriate number of cal/kg for each food item, as the dataset did not indicate if food was cooked or uncooked, or what parts of the food item were consumed. Using the United States Department of Agriculture calculator [16], I assumed that foods cooked and consumed in their entirety; e.g., the cal/kg for chicken was determined from the cal/kg in a cooked, whole chicken boiled without salt, as opposed to raw chicken thighs. For some foods where there was no information available, the nearest food substitution was used, e.g., rice noodles for *bihon* (rice vermicelli). Another assumption of this study is that the minimum recommended caloric intake for healthy, active adults is 2000 calories, although some studies suggest that this should be modified for Filipinos [17] and, presumably, other ethnicities.

The third and fourth indicators, absolute grams of protein and absolute grams of fat, were derived from the calories calculated. Using the standard conversions of one gram of protein equals four calories and one gram of fat equals nine calories, the absolute number of each macronutrient is the product of the amount of food (kg), the calories per kilogram in the food (cal/kg), the ratio of macronutrient to others in the food, and the conversion factor of calories to grams for the macronutrient (cal/g). I chose these metrics since one of the major malnutrition problems in the Philippines, as in many other developing countries, is a lack of adequate or disproportional amount of protein in the diet.

While I did not know a priori what type of distribution the annual per capita consumption of food might follow, power law analysis appeared to be a strong tool for analyzing the dataset since food systems may be thought of as complex systems that are diverse, connected, interdependent, and adaptive. Should the distribution of food consumption obey a power law, it would reveal important insights into managing food security and maintaining healthy diets across different populations. For instance, the concept of a system as "metastable" means that there may exist "one or more potential states... other than the extant state, to which the system could transition, given the realization of certain conditions" [18]—information that would be useful to policymakers on local, regional, and national levels interested in keeping a food system stable and secure. It may also be worthwhile to see if the concepts of phase transitions or tipping points may be applied within a social food system, e.g., if different foods can displace each other and change the dietary makeup of a population. Essentially, power law analysis can reveal features of food security and stability that traditional analyses of nutrition may have overlooked before.

11.3 Results

At first, the distributions of annual consumption of agricultural goods per capita across the four socioeconomic classes appear to obey Type I (Auerbach–Zipfian) power laws, as they have a one extremely large value, some mid-range values, and a heavy tail of small values (see Fig. 11.1). However, upon plotting the log–log values of rank and frequency (see Fig. 11.2), the resulting distributions display significant bending, particular in the lower half of the data. By choosing the dividing point to be the rank at which the bending appears to be at a maximum, the distribution can be split into

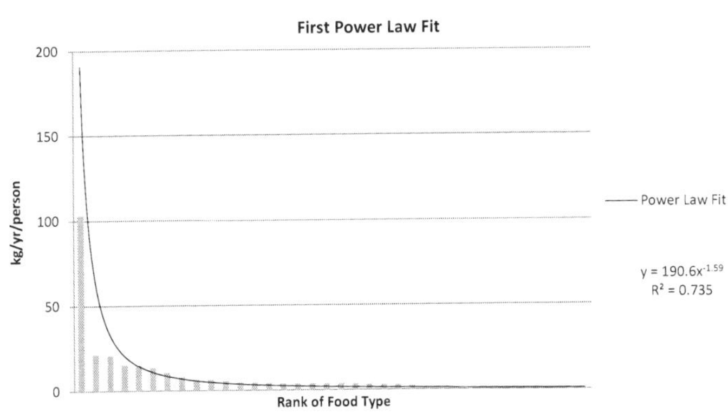

Figure 11.1 Power law fit for annual per capita NCR upper class consumption of foods (kg/yr/person).

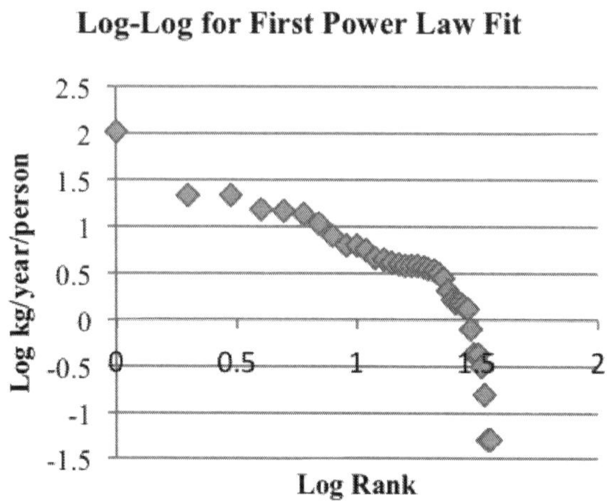

Figure 11.2 Log–log plot of annual per capita NCR upper class consumption of foods.

two regimes of food products. The first regime fits a Type I power law much more accurately than the original total distribution, as evident by an almost linear log–log plot (see Figs. 11.3 and 11.4), while the second regime appears to fit an exponential trend line more accurately, even though the log–log plot is not completely linear and displays bending of its own (see Figs. 11.5 and 11.6).

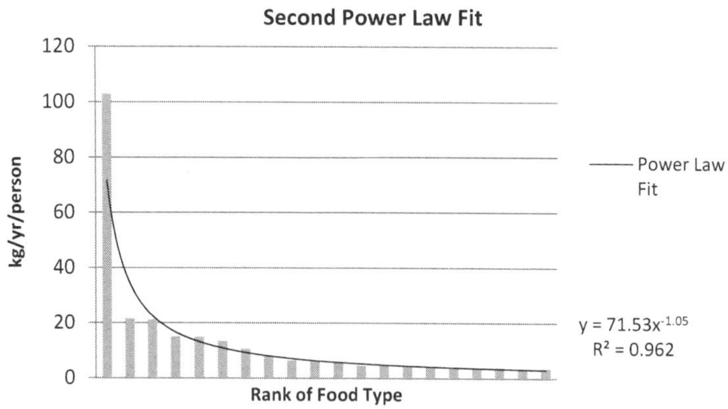

Figure 11.3 Power law fit for food regime I of annual per capita NCR upper class consumption of food (kg/yr/person).

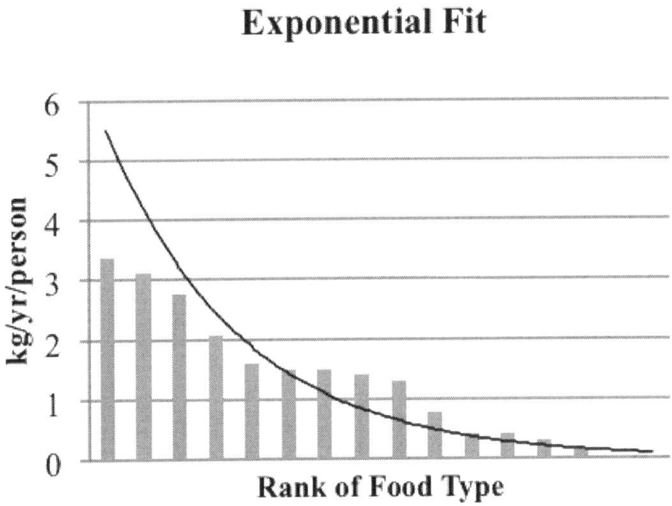

Figure 11.4 Exponential fit for food regime II of annual per capita NCR upper class consumption of food (kg/yr/person).

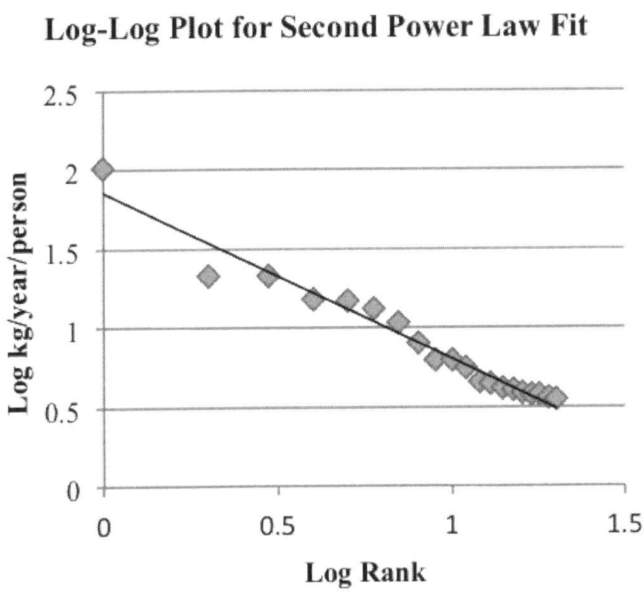

Figure 11.5 Log–log plot and linear regression of food regime I of annual per capita NCR upper class consumption.

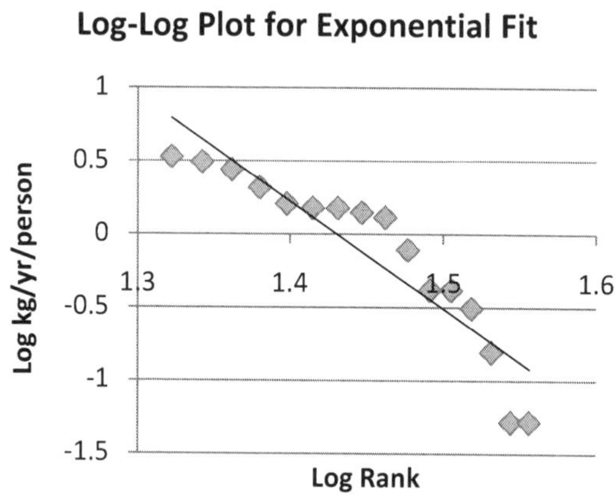

Figure 11.6 Log–Log Plot and linear regression of food regime II of annual per capita NCR upper class consumption.

As the wealth of the socioeconomic class decreases, the first regime distribution fits a power law less accurately (see Table 11.1). Within the upper class, there are more foods that can be classified as mid-range values, that is, values between the extremely large value and the long tail. However, as the classes decrease in status, consumption of some of these first-regime foods decreases such that it may be more accurate to classify them as small values. Simply put, the lower classes are maintaining one food that has a large extreme value (rice), but they are consuming less (non-rice) food overall.

Table 11.1 Comparison of power law fits and *R*-squared values across classes

Class	Power law fit	R2 value
Upper	$y = 71.538x^{-1.051}$	0.9628
Middle	$y = 59.015x^{-1.102}$	0.9554
Lower	$y = 51.978x^{-1.194}$	0.9333
Extremely lower	$y = 39.168x^{-1.159}$	0.8688

The ratio of macronutrients across classes deviates from the standard recommended daily allowance of 60% carbohydrates, 25% fats, and 15% protein (see Fig. 11.7). Even if one class approximates part of the recommended standard in one way, they all deviate from the standard macronutrient values. For instance, while the lower class most closely approximates the standard for carbohydrates compared to the other classes, the class is deficient in fat intake and has an increased protein intake. Furthermore, as classes decrease in social status, they increase in carbohydrate intake relative to fat and protein consumption. All classes consume more protein than the recommended standard, and more protein relative to fat, indicating that classes may be increasing protein to compensate for a lower fat intake.

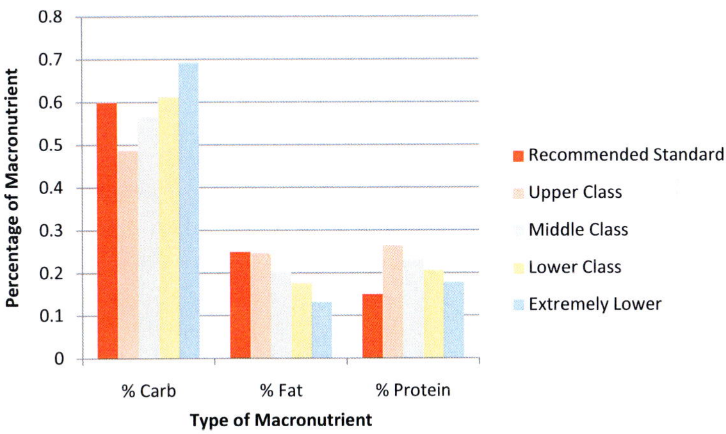

Figure 11.7 Proportions of macronutrients across socioeconomic classes.

The calculated value of absolute calories from the BAS dataset indicates a disparity between the dataset and official surveys of calorie consumption. However, it is first important to note that there are discrepancies between official surveys. FAO national surveys reveal that calorie consumption across the whole of the Philippines exceeds a standard diet of 2000 calories, but this contrasts with the Philippine Food and Nutrition Research Institute, Department of Science (FNRI-DOST) surveys that indicate that calorie consumption across the nation is much lower than the recommended values (see Table 11.2). There is a difference of over 500 calories between FAO 1995–1997 data and FNRI-DOST 1993–1994 data, 25% of the calories of a standard diet. Even though these are surveys of the entire country instead of just the NCR, data from the FNRI-DOST suggests that the average caloric intake from the population within the NCR should be very close to the average caloric intake of the whole country. The main conclusion to be drawn from these results is that there are discrepancies between official surveys, so it is quite possible that other surveys would also yield contradictory values.

Table 11.2 Comparison of FAO, FNRI-DOST, and BAS datasets across relevant years and regions

Data source	Region	Class	Year	Carb	Fat	Protein	Total calories
Standard				0.6	0.25	0.15	2000
FAO	National	All	1990–1992	0.738	0.168	0.094	2230
	National		1995–1997	0.736	0.207	0.093	2340
	National		2000–2002	0.724	0.183	0.093	2400
	National		2005–2007	0.732	0.174	0.094	2520
FNRI-DOST	National		1993–1994	0.732	0.150	0.119	1684
	NCR			n/a	n/a	0.127	1651
BAS	NCR	Upper	2008–2009	0.487	0.248	0.265	1337.40
	NCR	Mid.		0.566	0.203	0.232	1038.59
	NCR	Low		0.614	0.178	0.208	819.19
	NCR	Ex. Low		0.692	0.131	0.178	704.25

Note: Data obtained and calculated from FAO Statistical Yearbook 2010 and FAO Philippine Nutrition Profile [9].

Indeed, the calculated values from BAS show that all caloric intake of all socioeconomic classes falls well below the recommended daily allowances. Therefore, for this study, it is best to reevaluate the usefulness of the metric of absolute numbers of calories in determining malnutrition. A more useful metric would instead be the ratio of macronutrients across classes (as shown in Fig. 11.7).

Although absolute numbers of calories may not be as reliable a metric due to disparities within the data, the absolute numbers of grams of protein and the absolute numbers of grams of fat may still lend insight into malnutrition in the Philippines. Fat intake appears standard across classes: pork is the main contributor of fat to diets, as it is consumed more than three times as much as chicken. The range of fat contributors is limited to primarily animal fat. However, protein is more varied across classes (see Fig. 11.8). All classes have at least five main contributors of protein (foods with a yearly intake of more than 1000 grams), one of which is a non-traditional source. National Capital Region residents consume so much rice that, although it is primarily a source of carbohydrates, it also becomes a main source of protein.

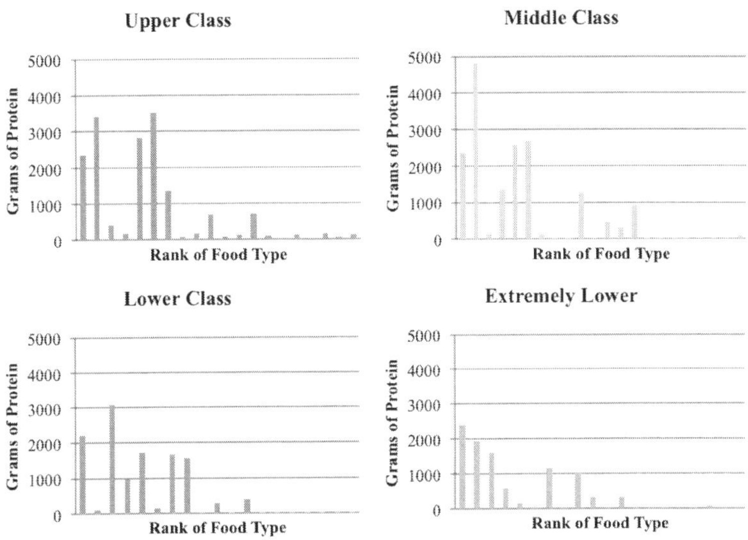

Figure 11.8 Protein consumption across class in grams per person per year (g/person/yr) as according to ranking of food in each class.

11.4 Discussion

If consumption of food follows a power law, this implies that consumption of food is a complex system. Foods may be considered as connected and interdependent entities; i.e., the consumption of one food occurs with consumption of others. Clearly documents such as recipes, menus, and cookbooks demonstrate that food consumption is diverse and not isolated, that certain foods are paired with others as the circumstances allow, and that these combinations of food evolve over time and adapt to cultural tastes and historical events. Those circumstances may be for reasons of desirability (taste, health) and accessibility (cost, physical proximity), two characteristics which may hold different meanings and interpretations across classes.

Across classes, the foods consumed fit into two different regimes as recorded in Table 11.3 (the regime split is designated by lighter colors in Regime II). I argue that items in "Food Regime I" are foods considered "essential," meaning that there is rarely an alternative food to replace them on an individual's plate. For instance, a person is either consuming rice or not; there is rarely a bread or pasta substitute that is considered culturally equivalent. If a recipe calls for pork, it is unlikely that chicken or fish will be used in its place. However, this quality for a given food item is not consistent across classes: in a sense, there is a "forced fluidity" or "directed fluidity" among the lower class.[d] Because the lower classes may have limited financial and cultural access to foods, they are compelled to be more resourceful and substitute food more frequently than a wealthier person. Because chicken cannot be purchased today, the alternative roundscad fish may be cooked and consumed instead.

Table 11.3 Rank and type of food consumed across socioeconomic classes

Rank	Upper	Middle	Lower	Extremely lower
1	Rice	Rice	Rice	Rice
2	Chicken	Chicken	Banana	Chicken
3	Pork	Banana	Chicken	Roundscad
4	anana	Pork	Pork	Pork
5	Milkfish	Tilapia	Roundscad	Pandesal
6	Tilapia	Milkfish	Pandesal	Banana
7	Beef	Pandesal	Tilapia	Tilapia
8	Mango, ripe	Mango, ripe	Milkfish	Eggplant
9	White potato	Roundscad	Eggplant	Milkfish

[d]I make the distinction between these two terms because "forced" may imply that the lower classes have no agency in determining what foods they eat, while "directed" implies that there are certain limitations that place their access to foods within a smaller conduit, but the population still maintains a certain amount of choice given their financial constraints.

10	Chicken egg	Eggplant	Inst. noodles	Inst. noodles
11	Papaya, ripe	Ampalaya	Chayote	Chayote
12	Onion	Chicken egg	Chicken egg	Chicken egg
13	Roundscad	Inst. noodles	Mango, ripe	Tomato
14	Inst. noodles	Beef	Tomato	Onion
15	Ampalaya	Chayote	Onion	Stringbeans
16	Squash	Pineapple	Ampalaya	Mango, ripe
17	Pineapple	Onion	Cabbage	Ampalaya
18	Tomato	Cabbage	Stringbeans	Garlic
19	Eggplant	Squash	Squash	Squash
20	Cabbage	Tomato	Garlic	Cabbage
21	Pandesal	Stringbeans	Pechay	Pechay
22	Chayote	Garlic	Okra	Mongo
23	Garlic	Pechay	Beef	Camote
24	Stringbeans	White potato	Mongo	White potato
25	Pechay	Papaya, ripe	White potato	Okra
26	Okra	Okra	Camote	Beef
27	Pasta	Camote	Habitchuelas	Cassava
28	Mongo	Mongo	Pineapple	Pineapple
29	Canton	Bihon	Bihon	Habitchuelas
30	Habitchuelas	Habitchuelas	Cassava	Papaya, ripe
31	Bihon	Canton	Papaya, ripe	Bihon
32	Gabi	Pasta	Corn	Pasta
33	Camote	Cassava	Pasta	Canton
34	Corn	Milk	Canton	Gabi
35	Cassava	Corn	Gabi	
36	Milk	Gabi	Carabeef	
37		Fresh Miki	Fresh Miki	

On the other hand, the upper class may be thought of as having a "free fluidity," which means they have greater financial means to purchase a wider variety of foods. They may often choose a food that is more expensive, subjectively tastier, and of "higher status." However, because they often have the ability to choose a higher quality food, their selections solidify into an array of foods that are rarely substituted for each other because necessity does not demand it. In some regards, the claim might be made that the foods consumed by the upper class represent the idealized diet (not necessarily healthiest) for Filipinos within the NCR, should they have the budget for it.

Foods in Regime II do not follow a power law distribution, and instead appear to be consumed in amounts roughly equal to each other. This suggests that items in this part of the distribution are "substitutable" foods, meaning they are alternatives to each other, or less accessible or desirable than foods in the Regime I. For instance, many grain- or cereal-based carbohydrate substitutes such as pasta, *bihon*, and *canton* are in this regime, likely to be potential substitutes for rice, but more likely to be used as equivalent substitutes for each other. In a sense, they can be thought of as competing for the same cultural space on a plate, but factors of finance, taste, physical access, and nutrition will determine which item is eaten.

Another revealing aspect of Regime II is the accessibility and desirability of foods. For instance, beef and pineapple are items found in Regime I for the upper and middle class, yet they exist in Regime II for the lower and extremely lower class. I suggest that foods that are both in Regime I for the upper and middle class and in Regime II for the lower and extremely lower class are desirable across classes, but not accessible to all. Using the upper class consumption as the standard diet, this would imply that desirability and accessibility are inversely proportional to each other as classes decrease in status. Foods that are found within Regime I across classes are both desirable and accessible, and foods found within Regime II across classes are not as desirable but still accessible.

As for the other indicators of malnutrition, the number of calories consumed and the grams of macronutrients derived from the calories suggest that all classes in the NCR are undernourished, though these findings do not correspond with data from other official surveys. From the BAS dataset, it would seem that obesity is not a problem in the NCR, while quantitative and qualitative data suggests otherwise. Therefore, for this dataset, it is more useful to view relative numbers of calories and macronutrients and compare them across classes. Furthermore, the fact that there is significant bending in the power laws as well as their log–log plots further suggests that smaller values of consumed foods are not recorded, which may also account for the low count of calorie intake across classes. Low calorie intake may also be attributed to the fact that many demographic groups are masked in the collection of the data, i.e., there are no distinctions in the data for gender, age, health, or pregnancy status. On the other hand, this data does support the issue of PEM, albeit non-intuitively. Filipinos consume more than the recommended amount of protein, yet not enough fat, which in turn sets their bodies to consume protein instead of fat, thereby losing fat-soluble vitamins (hence vitamin A and other micronutrient deficiencies) and damaging healthy cell growth, among other debilitating consequences.

While there are many implications that can be derived from this data, it is critical to highlight the role of rice in the Philippine diet. The cultural food system of Filipinos in the NCR appears stable because, in general, rice substitutions are restricted to Regime II and are not perturbing the amount of rice consumed (with the possible exception of instant noodles). Rice accounts for between one-third and two-thirds of the Filipino diet in terms of calories, and the amount of rice consumed increases as the socioeconomic status of the class decreases. People in the lower classes may be consuming rice for several reasons, some of which may be because of its high accessibility and lower cost. So much rice is consumed that it provides a significant amount of protein, particularly to people in the lower class.

However, the system would become metastable if an event were to displace rice from the first rank. Should a natural famine strike, or should even fear of shortage and price inflation cause rationing and a famine, the system would likely transition into an unstable state with dire consequences across the population, but particularly for the lower classes, much like the 1943 Bengali famine [2]. In a more optimistic light, the system might also become metastable should a new food or crop be introduced to replace rice, which may stabilize the system or offer more diverse choices for people's diets. However, this data suggests that new foods (pasta, noodles) and crops (corn, cassava) were introduced to replace rice, yet as they remain in Regime II, the locals have not adopted them as a significant source of food. These foods have the potential to be tipping points for the system, but they have not been adopted and adapted by the local population.

Instant noodles are the exception to these competing rice substitutes and have potential to be the commodity, which causes the system to have a phase transition. Around the time of the release of this dataset, an article in the Philippine Inquirer denounced the rise of instant noodles as it replaced other foods and the negative impact it had on health, particularly that of children [19]. People are consuming instant noodles since they are a source of cheap, tasty food in comparison to increasingly expensive rice and other starch substitutes. The fact that many have been fortified with vitamins may mislead consumers into thinking that the noodles provide a well-balanced meal. Consequently, the food system is transitioning toward stability but not necessarily toward higher quality nutrition.

11.5 Future Research Possibilities

There is rich potential for future expansion and follow-up studies from this data and from the other datasets included in the BAS general survey. Distributions of consumption could be compared across time, regions, or different types of food. Furthermore, food could be analyzed in terms of its micronutrients. As vitamin A and iron deficiencies are serious issues in the Philippines [20], the proportions of vitamins and minerals could be calculated and compared across classes for indicators of malnutrition. Foods could be categorized differently, e.g., by comparing meat consumption to vegetable consumption or by analyzing green leafy vegetables intake compared to starchy vegetable intake. It should be tested across regions and classes if power laws are still common in the distributions of other sets. This kind of analysis may improve recommendations as to what types of food and amounts of food that should be modified to improve Filipinos' diets.

As data is available for additional regions of the Philippines, it would be useful to replicate this data analysis and determine if power laws appear in other parts of the country, or if the power law distribution is unique only to the NCR. What are the implications of finding a power law in urban areas of the country versus rural regions? Might a power law indicate different levels of malnutrition, or could it reveal attitudes toward availability and desirability of food? Furthermore, cross-country comparisons would also be beneficial for studying malnutrition beyond the Philippines. These comparisons could then be correlated with other metrics, such as the Global Hunger Index, to form a more robust understanding of consumption and hunger in the world.

Network science provides another useful approach for determining how people consume their food. By surveying what people eat, a researcher could create a dietary network of foods that are commonly eaten together, perhaps assigning weights to links and nodes such as "amount of food eaten," "type of food," "caloric value," or "perceived cultural status." If there are clusters of foods that people usually eat, or foods that have a greater centrality than others, this information would be helpful for nutritional policy leaders in determining how to best target their health campaigns, perhaps by focusing on how to improve these foods' nutritional value or how to enhance or limit their distribution to the population.

For practical and policy purposes, ethnographic research would complement power law analysis or network science research by providing context for the quantitative data. In fact, by conducting power law analysis prior to fieldwork, a social scientist could use their findings to form or predict their hypotheses on malnutrition. Upon entering the field, conducting interviews, and engaging participant observation, the researcher could determine on the micro-level how people identify foods as accessible and desirable, and if this is in line with the power law analysis.

To strengthen this study, it would be beneficial to cross-reference this data with economic and cultural references. For instance, checking the 2008–2009 prices on these commodities would be helpful in determining how much financial access classes may have actually had to these foods. Similarly, it would be helpful to reinforce the ideas of desirability and accessibility with cultural observations of food consumed across classes. While the assumption of this study has been that the upper class has an "ideal diet," it is quite possible that the lower classes have an ideal diet within their own sphere of influence. For instance, since the Spanish colonization of the Philippines introduced a host of new crops and recipes to the residents, upper class foods have a history embedded in Spanish cuisine—people of the upper class are more likely to consume costly foods, like beef, and scarcer, imported foods, like pineapple. Lower class people would have been more likely to consume indigenous foods and rely heavily on fish. Even with greater accessibility to food, because of time and tradition, it may be that people in the lower class would desire fish and chayote more than beef and pineapple.

It is not conclusive from this data if following a power law makes a Filipino's diet "ideal." While the distribution of the upper class most clearly follows a power law and for the purposes of this study has been labeled the "ideal diet," it is only ideal within its own cultural space. That is, the diet is ideal to Filipinos with enough money and accessibility to those types of foods. It is not necessarily the most nutritious diet to follow. However, the fact that the distributions follow power laws indicates measures of inequality across classes and has implications for metastability. For instance, the fact that the quality of the fit of power law decreases as status of class increases indicates that the mid-range values of the lower classes are missing—the lower class is eating less and may have less accessibility to certain types of food. The focus on rice also indicates that the system could easily become metastable under famine, price inflations, or introduction of a new foodstuff. Therefore, greater food security may be achieved as more diversity is introduced into the diet, thereby inducing more stability into the system. However, it is equally important that the diverse foods being introduced also be healthful and dense in nutrients, so that the food system is stable as well as wholesome. If researchers viewed malnutrition and food consumption as complex adaptive systems, policymakers could then apply system properties across different cultures and countries to reduce malnutrition and increase health across various populations.

11.6 Summary

Because of its unique historical status and the availability of data, I chose the Philippines as a case study for power law analysis of malnutrition across socioeconomic classes based on consumption of foods. I created a composite metric of four criteria in order to measure malnutrition across classes. The major finding of this research is that the data from this survey creates a distribution that can be split into a power law and an exponential distribution, creating two regimes of food. Examining the regimes of foods as they pertain to an upper and lower class yields insight into the desirability and accessibility of certain foods. The fact that the first regime follows a power law also has implications for the stability and food security of a nation's social food system. It is highly recommended that in order to achieve stability and offset events such as famine or price inflation, the food system should be diversified and not rely solely on one type of food. Furthermore, as the food system is diversified, policymakers should also take into account the nutritional status of the food, moving the entire system toward stability as well as health. More exploration should be made before concluding that a power law of food consumption is "ideal."

References

1. *Malnutrition—PubMed Health* (2009, May 12). Retrieved May 10, 2011, from PubMed Health: http://www.ncbi.nlm.nih.gov/pubmedhealth/PMH0001441/.
2. Bodley, J. (2008). Malnutrition and the Evolution of Food Systems. In *Anthropology and Contemporary Human Problems* (5th Edition ed.). New York: Altamira Press.
3. The Spectrum of Malnutrition (2002). Rome, Italy: World Food Summit.
4. Gillespie, S., and Haddad, L. (2001). *Attacking the Double Burden of Malnutrition in Asia and the Pacific.* Washington, D.C.: International Food Policy Research Institute.
5. von Grebmer, K., and Ruel, M. T. (2010). *2010 Global Hunger Index.* Concern Worldwide and Weithungerhilfe. Washington, D.C.: International Food Policy Research Institute.
6. Agoncillo, T. (1990). *History of the Filipino People* (8th Edition ed.). Manila: Garotech Publishing.
7. *Background Notes: Philippines* (2010, October 29). Retrieved May 10, 2011, from U.S. Department of State: http://www.state.gov/r/pa/ei/bgn/2794.htm.
8. von Braun, J., and McComb, J. (1993). *Urban Food Insecurity and Malnutrition in Developing Countries: Trends, Policies, and Research Implications.* Washington, D.C.: International Food Policy Research Institute.

9. Bayani, E., and Marchesich, R. (2010). *Nutrition Country Profiles: Philippines Summary*. Retrieved May 10, 2011, from FAO Nutrition and Consumer Protection: http://www.fao.org/ag/agn/nutrition/phl_en.stm.
10. Adair, L. (2004, Aug 8). Dramatic Rise in Overweight and Obesity in Adult Filipino Women and Risk of Hypertension. *Obesity Research*, 1335–1341.
11. Bersier, L.-F. (2007). A History of the Study of Ecological Networks. In *Biological Networks* (Ed. Kepes, F.), New Jersey: World Scientific.
12. Banasek-Richter, C., and Bersier, L.-F. E. (2009). Complexity in Quantitative Food Webs. *Ecology*, **90**(6), 1470–1477.
13. Rossberg, A. G., Farnsworth, K. D., Satoh, K., and Pinnegar, J. K. (2011). Universal power-law diet partitioning by marine fish and squid with surprising stability-diversity implications. *Proceeding of the Royal Society B.*, **278**, 1617–1625.
14. Kinouchi, O., and Diez-Garcia, R. W. (2008). The Non-Equilibrium Nature of Culinary Evolution. *New Journal of Physics*.
15. Whitney, E., and Rolfes, S. (2010). *Understanding Nutrition* (12 ed.). Belmont, CA: Wadsworth Publishing.
16. *Nutrition Data*. (n.d.). Retrieved May 10, 2011, from NutrtionData.com: http://nutritiondata.self.com/.
17. Cabrera, I. (1998). Philippine Recommended Dietary Allowances: Recent Developments and Future Plans. *Nutrition Reviews*, **56**(4), 21–24.
18. Cioffi-Revilla, C. Analyzing Power Laws in the Social Sciences: A Primer. In *Power Laws and Non-Equilibrium Distributions of Complexity in the Social Sciences* (ed.Cioffi-Revilla, C.), in publication.
19. Bolido, L. (2009, June 30). *Instant Noodles are Not Balanced Meals.* Retrieved May 10, 2011, from Philippine Daily Inquirer: http://showbizandstyle.inquirer.net/lifestyle/lifestyle/view/20090630-213201/Instant-noodles-are-not-balanced-meals.
20. Darnton-Hill, I., and Webb, P. (2005). Micronutrient Deficiencies and Gender: Social and Economic Costs. *The American Journal of Clinical Nutrition*, 1198–1205.

Chapter 12

Exploiting the Synergy between the Micro, Meso, and Macro Levels in a Complex System: Bringing to Life an Ancient Urban Center

Thaer W. Jayyousi[a] **and Robert G. Reynolds**[b]

[a]*Department of Computer Science, Wayne State University, 3rd Floor, 5057 Woodward Avenue, Detroit, MI 48202, USA*
[b]*Department of Computer Science, Wayne State University, 14th Floor Suite, 14001, 5057 Woodward Avenue, Detroit, MI 48202, USA*

al6854@wayne.edu, reynolds@cs.wayne.edu

In this chapter, we describe the construction of agent-based models of emergent ancient urban centers using techniques from computational intelligence, agent-based modeling, complex systems, data mining, and existing archaeological data from the prehistoric urban center, Monte Albán. This application is selected because of its importance in understanding the emergence of modern economic and political systems. The use of agent-based techniques facilitates the migration of results and investigations between levels, macro, meso, and micro. We examine different spatial levels starting at macro level and then take its questions to the micro level thru the meso level to answer our questions, and then bring these answers back to the macro level.

Complexity and the Human Experience: Modeling Complexity in the Humanities and Social Sciences
Edited by Paul A. Youngman and Mirsad Hadzikadic

ISBN 978-981-4463-26-3 (Hardcover), 978-981-4463-27-0 (eBook)
www.panstanford.com

12.1 Introduction

We take urban centers for granted, but until just a few thousand years ago they did not exist. They have evolved over the last thousand years—a relatively short period of time. How did these early urban centers evolve? Why did they evolve? What did they look like? We know how cities look in the present day. We know how people work in the cities of today but what about these early cities? How did they develop? Are there any differences between early cities and today's cities? Can some of the issues that emerge be useful in understanding the nature of modern cities? This chapter investigates the use of agent-based techniques and virtual world technologies in order to answer the above questions. We use tools from complex systems and computer game technology as vehicles for the incubation and development of agent-based models of early urban formation. Our focus is on one prehistoric urban center, Monte Albán, which was the ancient capital of the Zapotec civilization and one of the first urban centers in the Americas. We selected this application because of its importance in understanding the emergence of modern economic and political systems and the availability of an extensive data set collected by archaeologists studying the site.

In order to address the questions raised above we employ techniques from computational intelligence, agent-based modeling, complex systems, and data mining to analyze the data set produced by surveying the Monte Albán in light of the questions posed above. Given the temporal and spatial expanse of the site, Kowalewski [1] suggests that the answers to the above questions will require "the examination of trends and events proper to a number of different temporal levels, first separately and then in relationship to one another." We then address these questions at the three different spatial scales: the macro level (the site), the meso level (neighborhoods or barrios), and the micro level (terraces and individual residences). The findings produced at one level can be applied to other levels to address related questions there.

More specifically, complex systems operate on a number of temporal and spatial levels. It is frequently the case that the data collection process is concentrated on one level or another. Questions or hypotheses posed at the macro level may not have sufficient data to test them at that level. Thus, it is important to be able to traverse levels such that questions asked on the macro level with sparse data can be re-expressed at the micro level with sufficient data. The results of the analysis at the micro level can then be re-expressed at the macro level as well as other levels that may utilize the results. In this chapter, we propose techniques that facilitate the movement of our research focus from one level to another and back again. As a result, we are able to build a model that will be the basis for our virtual world.

12.2 Cultural Algorithms

Cultural algorithms simulate the cultural evolution process, conveying a broader learning and evolution than simple biological evolution such as genetic algorithms. They can be used in complex multi-agent systems to provide valuable simulations of learning procedures. The cultural algorithm is a socially motivated algorithm [2], and was developed by Reynolds 1979 [3–5] as an extension of Holland's genetic algorithms to the modeling of cultural evolution and social decision-making.

The cultural algorithm is a dual-inheritance system that maintains two search spaces: the population space, which represents a collection of individuals embedded in the culture component, and a belief space, which represents a cultural component. While the population space represents the individuals in terms of a genotypic and/or phenotypic level, the belief space models the cultural information about the population. Both the population and belief spaces evolve in parallel, with both influencing one another using two communication channels (Fig. 12.1).

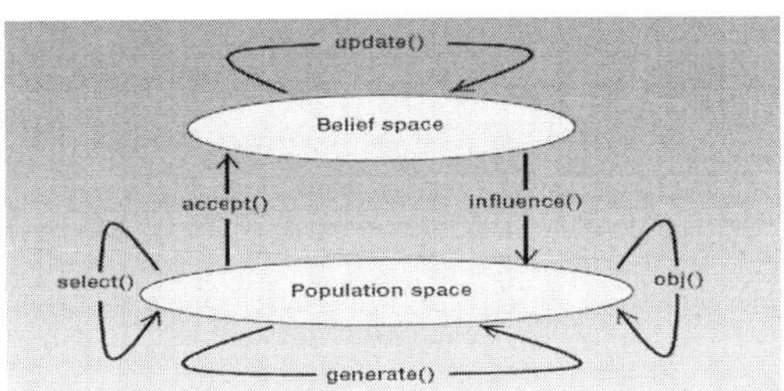

Figure 12.1 The cultural algorithm has three major components: a population space, a belief space, and a protocol that describes how the first two components exchange knowledge. The population space can support any population based computational model, such as genetic algorithms or evolutionary programming.

In multi-agent problem solving, agents work together or compete against each other to achieve a goal. To be effective, the agents need to interact, and they need to behave cooperatively and/or competitively in order to accomplish an objective. The central issue is how such cooperation or competition can be best established. Cultural algorithms are particularly useful in this regard [6,7]. Since cultural algorithms by definition contain two components, the belief space and the population space, it has the ability to support cooperation and competition within each component and between components.

12.3 The Emergence of Monte Albán: An Example Approach

12.3.1 City Morphologies: Hypothetical Models of Urban Growth

Cities such as Monte Albán are one of the most "spectacular settlement types of pre-Columbian Mesoamerica" [8]. These prehistoric urban centers were both the product of social changes and the platform on which future changes would take place. These archaic cities can be described in terms of the following dimensions:

- The size of the city either in terms of population or area covered by the city.
- The location of the city in terms of its environment: on a river, a coast, a mountaintop, or another location.
- The function of the city in terms of ceremonial, commercial, defensive, or administrative activities.
- The position of the city in a settlement hierarchy.
- The morphology or form of the city. The form that a city takes can be influenced by all of the variables above.
- The basic framework for our approach revolves around the form of a city because it is a reflection of the other variables above.

Doxiadis states that a city contains four basic parts [8]. They are

(1) Homogeneous parts such large multifamily living quarters.
(2) A central part or plaza.
(3) The circulatory part or road system.
(4) A special area such as a marketplace.

Not all of these parts are necessarily found in all cities, but they reflect features that are likely to be planned parts of the morphology or shape of a city.

The morphology of a city also has unplanned components that reflect its growth or decline over time. There are several models developed within the last half century to describe these emergent shapes in terms of modern cities. These models include the following:

(1) *The concentric zone model* based upon the growth of modern cities, with a city center surrounded by concentric zones of activity.
(2) *The sector model.* In this model, differences in land use near the center are preserved and "fan out" like slices of pie as the city grows. These differences can reflect residential, economic, and civic ceremonial differences.
(3) *The concentric-sector model.* This model combines both the concentric and sector models where activities are organized as either sectors or concentric zones of activity.
(4) *The multiple nuclei model.* This is an extension of the concentric model in which there are multiple nuclei in which activities are organized around several concentric zones.

12.3.2 The Emergence of Monte Albán

The archaeological site of Monte Albán is situated in Mexico's Valley of Oaxaca, located in central Mexico. Figure 12.2 shows the basic physical layout of the site, with its central plaza located to the south of the flat hilltop 400 meters above the valley floor. Table 12.1 gives all of the relevant periods of social evolution in the valley [9]. Tierras Largas marks the beginning of early village settlement there. The state emerged at Monte Albán in period Monte Albán Ic. The valley came under control of the state by Monte Albán IIIb, and Monte Albán IIIa signaled the decline of the state and its succession by a collection of city-states localized in different parts of the valley. The phases as described there represent uneven slices through time and are defined in terms of the pottery found at Monte Albán. Each phase change represents a change in the predominant style of pottery. The chronology of the phases was determined by radiocarbon dating the pottery of each style [9].

Figure 12.2 Monte Albán.

Table 12.1 The basic occupational phases of the Valley

Period	Approximate date
Tierras Largas	1400–1150 BC
San Jose	1150–850 BC
Guadalupe	850–700 BC
Rosario	700–500 BC
Monte Albán Ia	500–300 BC
Monte Albán Ic	300–150/100 BC
Monte Albán II	150/100 BC–AD 200
Monte Albán IIIa	AD 200–500
Monte Albán IIIb	AD 500–700/750
Monte Albán IV	AD 700/750–1000
Monte Albán V	AD 1000–1521

Monte Albán is distinguished by a central plaza and road system. According to Blanton [9], the site is broken up physically into neighborhoods defined by these ancient roads, defensive walls, and natural barriers such as barrancas. Blanton calls these neighborhoods "barrios." These barrios are illustrated in Fig. 12.3.

Some of these barrios may reveal occupational sectors while others might represent multiple nuclei of occupation over time. In the same figure, a white box in barrio 2 represents major part of the city, the Main Plaza.

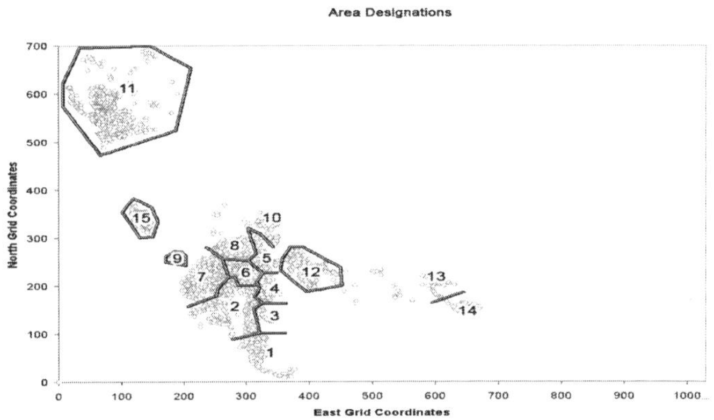

Figure 12.3 Map of Monte Albán showing the barrio neighborhood designations of Blanton. The white area in the middle of barrio 2 presents the Main Plaza of the city [16].

Another important part of Monte Albán is its road network; Fig. 12.4 shows majors roads in the city.

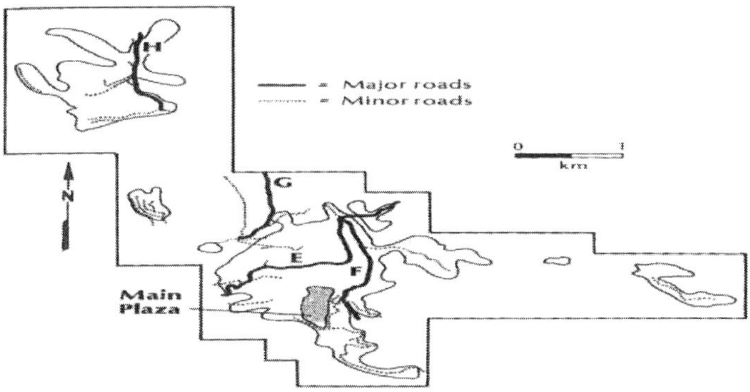

Figure 12.4 Major road networks at Monte Albán.

12.3.3 The Monte Albán Dataset

In early 1970s, the National Science Foundation supported an extensive survey of the Oaxaca Valley along with the site of Monte

Albán, the Valley of Oaxaca Settlement Project [1]. In the first phase of the Valley of Oaxaca Settlement Pattern Project a detailed mapping and surface collection of the region's major archaeological site, Monte Albán was conducted.

Archaeological surveys of the site plus excavations have produced a large data set consisting of 2073 occupational terraces (units, houses) that make up the Monte Albán occupation. Each part of the site is designated as a terrace because the site is situated on a hill and occupational terraces are carved into the sides of the hill. Each terrace is described in terms of several hundred environmental and cultural variables which come from an intensive archaeological survey undertaken as part of the Valley of Oaxaca Settlement Pattern Project [1]. Table 12.2 gives several example terraces along with some of their environmental variables.

Table 12.2 The environmental variables for four example terraces at the site are presented

Attributes	Instance 20	Instance 1234	Instance 272	Instance 260
Location	Monte Albán	Monte Albán	Monte Albán	Monte Albán
North grid coordinate	189	183	175	171
East grid coordinate	340	339	311	308
Elevation	375	375	400	400
Topography	Sloped	Sloped	Near flat	Near flat
Soil type	1	1	1	1
Soil depth	0	0	0	0
Silting	None	None	None	None
Presence of a spring	Absent	Absent	Absent	Absent
Baranca or wash adjacent	Absent	Absent	Absent	Absent
Type of vegetation	Grass and brush	Grass and brush	Grass and brush	Grass and brush
Vegetation abundance	Moderate	Moderate	Moderate	Moderate

Special resources	Quarryable stone	Quarryable stone	None	None
Distance from road	Close	Close	Close	Close
Occupied in Ia period	No	Yes	No	Yes

Note: Two of the terraces were occupied initially and two were not.

12.4 Exploiting Multiple Spatial Levels: Macro, Meso, and Micro

12.4.1 Introduction

Focusing now on the temporal scale, we investigate the structural/functional organization of the site during its first 200 years, Monte Albán Ia from 500 BC to 300 BC. However, the terraces associated with that time period were not all occupied over the same length of time. Some were occupied first while others occupied later as the more desirable sites filled up.

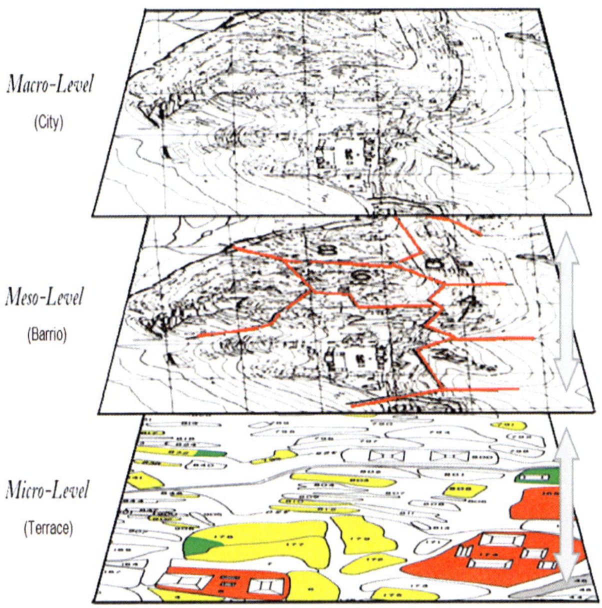

Figure 12.5 From macro to meso and micro.

We will look at three basic spatial levels here (Fig. 12.5):

Macro: Corresponds to the entire urban region. In this level, we investigate the physical morphology of the city and see if the morphology of the city is similar to one of the module we discussed early.

Meso: Corresponds to the Barrio level (neighborhoods or sub-divisions). It turns out that when the archeologists surveyed the city, they found evidence of sub-divisions within the area of the city. These sub-divisions are called barrios. An interesting question is: were these barrios operational during the first phase of occupation of the city?

Micro: Corresponds to the Terrace level, individual occupational sites. Terraces contain the actual houses where people lived at that time. We need to find out what were going in these terraces using the extensive data collected by the archeologists.

At the Macro level, we are trying to compare the occupational structure to the various city models presented earlier. At the Meso level, we are trying to compare the barrios in terms of their functionality. At the Micro level we are trying to compare individual terraces. The question then is, what data do these terraces have in common?

It is typical of archaeological data that it is heavily weighted towards the micro level. While each terrace can be described in terms of hundreds of variables, any one terrace has a much smaller subset. Thus, comparing terraces across the site (macro level) can only be done in terms of variables that all terraces have in common, such as environmental variables. However, it is difficult to compare terraces at this level of detail in order to answer our questions because only a small amount of data, primarily the relative location of terraces to environment and built feature. At the next level down, barrios can be compared using the environmental variables plus additional cultural variables that can be accumulated and compared across barrios. Finally, at the terrace level we can compare terraces in a barrio with each other using all of the variables associated with a terrace.

We start by asking questions at the macro level about the morphology of the city and then re-express these questions in terms of the barrios. Then, we re-express the barrio comparisons in terms of the terraces that make up each barrio at the micro level. Answers that emerge at the terrace level can be re-expressed at the barrio level and then finally at the macro level, the level of the entire site.

12.4.2 Macro Level (The City)

At the Macro level typical approaches are based on data commonly available at all locations. At this level we seek to determine what these terraces have in common in terms of global position and relative position to common features. In other words, we are trying to find out whether these terraces were occupied at a given time so as to reconstruct the overall morphology of the city as it emerged. We wish to compare that morphology with the current set of available models of the urban landscape discussed earlier.

Figure 12.6 shows plots of all terraces occupied in Monte Albán Ia. The red dots represent terraces that are occupied during that phase and the triangle in the middle of the plot presents the Main Plaza, a major focal point of the city. Environmental variables, distance from Main Plaza and distance from road were used in our data at the macro level to analyze the site.

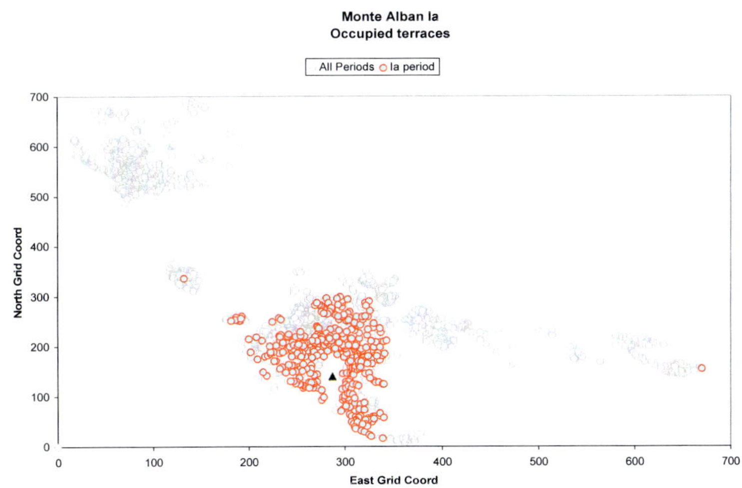

Figure 12.6 Terraces occupied during Monte Albán phase Ia, the triangle in the middle of the plot presents the Main Plaza.

Decision trees is a classic machine learning technique developed by Quinlan [10] in the mid 1980s. We have successfully used decision trees used in our research to characterize archaeological sites at the macro level [11]. We used environmental data (Table 12.3) since that is the only data available across all terraces. Our focus in this level was in predicting terrace occupation for period Ia and to compare predictions with actual occupation. Figure 12.7 shows a plot of a decision tree produced by applying J48 (C4.5) Decision Tree algorithm on environmental variables and Table 12.4 shows the run information for that tree.

Table 12.3 Distribution of the environmental variables for Monte Albán Ia

Variable	Values	Number of terraces	Min	Max	Mean	Standard deviation
Elevation	Numeric	390	125	400	348.462	45.992
Topography	Near_flat	161				
	Sloped	203				
	Flat_ridgetop	6				
	Hilltop	20				
Barranca or wash	Absent	264				
	Present	125				
Vegetation	Grass_and_brush	222				
	Grass_only	3				
	Cultivated	64				
	None	1				
Vegetation abundance	Moderate	357				
	Sparse	21				
	Heave	10				
	None	2				
Road distance	Far	100				
	Directly_adjacent	101				
	Close	88				
Distance from Main Plaza	Grid unit	390	13	383	76.531	40.053

Table 12.4 Run information for desired/undesired location in Monte Albán period Ia

Scheme	WEKA classifiers J48
Instances	776 instances; 388 instances from Monte Albán Ia period and another 388 instances from Monte Albán other periods extracted randomly
Test option	10-fold cross validation
Class	Ia Yes, Ia No
Correctly classified instances	640 (82.47%)
Incorrectly classified instances	136 (17.53%)
TP rate (yes)	334 (0.86)
TP rate (no)	306 (0.79)
FP rate (yes)	82 (0.21)
FP rate (no)	54 (0.14)
Minimum number of objects	5

The overall predictability for desired and undesired locations (yes/no) is 82% while the prediction for desired location (yes) is 86%. From Fig. 12.7, it is clear that the distance to the Main Plaza is the clearest organizing influence. Road distance is only a factor relative to topography and elevation. Terraces on nearly flat topography at higher elevations are more likely to be occupied if they are far from the road. On the other hand, those terraces at lower elevations that are farther from the Main Plaza are more likely to be occupied if they are close to the road.

Figure 12.8 plots the rules for the desirable terrace. The rules are spatially localized and tend to follow barrio subdivisions. The red terraces are desirable according to these rules. All other recorded terraces are represented in gray. The areas near to the Main Plaza along with the north-south road, which ends at the north entrance, are all highly desirable according to these rules. However, since the site is just becoming settled there may not be enough individuals to occupy all of the desirable terraces. Figure 12.9 on the other hand shows the terraces that are predicted to be unoccupied by the rules. They are, as expected, farther from the main plaza and lower in elevation. Figure 12.10 gives a plot of the terraces that are predicted to be desirable. They are given in red. Terraces that are predicted to be occupied, but are not, are in blue. These terraces are, in general, farther from the main plaza. This suggests that the population at this time was not sufficient to occupy all of the desirable terraces. Thus, the urban center was still in the emergent and "building" phase.

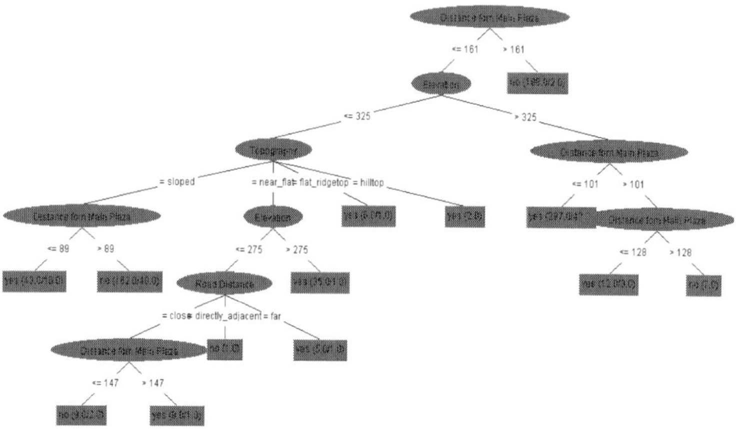

Figure 12.7 J48 decision tree for desired/undesired location in Monte Albán period Ia.

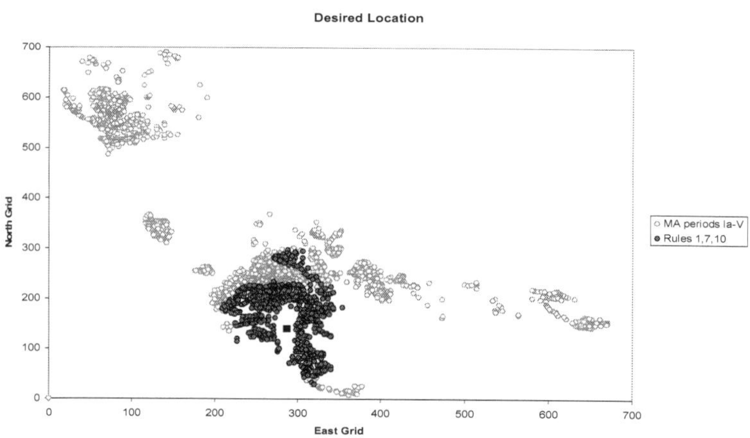

Figure 12.8 Plot of desired location rules applied to terraces occupied over Monte Albán Ia–V periods. The black square presents the Main Plaza. Red dots represent desirable locations according to the rules, and gray terraces represent the rest of the terraces occupied during Monte Albán periods Ia–V.

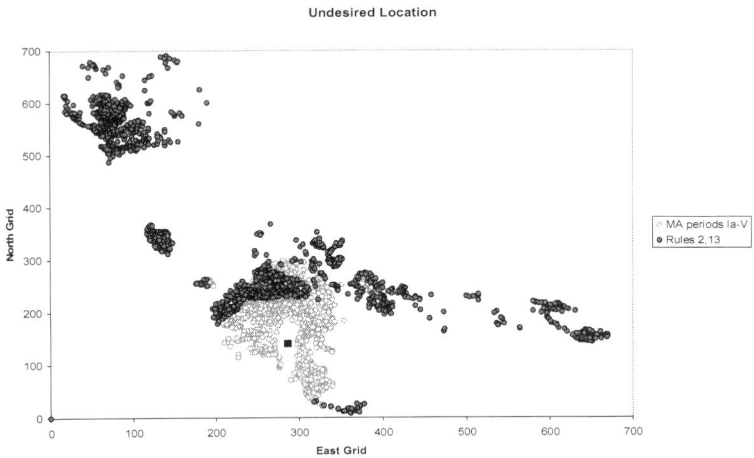

Figure 12.9 Plot of undesirable location rules for all occupied terraces from Monte Albán Ia–V periods. The black square represents the Main Plaza, the red terraces represent desirable locations, and the gray represent the rest of terraces occupied during Monte Albán periods Ia–V.

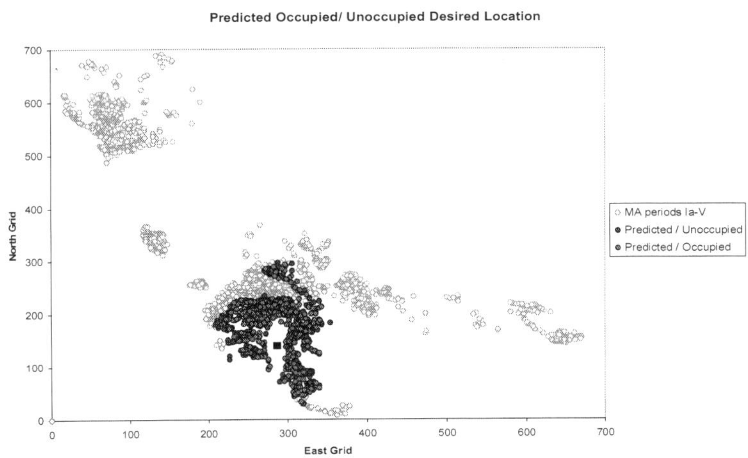

Figure 12.10 Predicted occupied/unoccupied desirable applied to Monte Albán Ia–V periods. The black square represents the Main Plaza, the red terraces represent the predicted and occupied desirable locations, while the blue represents the predicted but unoccupied locations, and the gray present the terraces occupied during Monte Albán periods Ic–V.

12.4.3 Meso Level (The Barrios)

Based on the previous results, it is worth examining the data at the barrio level since it appears the terraces are organized around these barrios. At this level we use multi-dimensional scaling to observe the similarities between the sub-divisions. We use material content to compare the barrios rather than environmental variables to see how similar or different these sub-divisions are. We plot these barrios within multi-dimensional space and observe how these barrios are closed to each other in terms of a few dimensions.

The macro level analysis suggests that the terraces' contents reflect their position within a barrio. While one cannot date the physical structure of the barrios, it appears that even in the first phase of occupation their influence was present. We use multi-dimensional scaling to aggregate terrace content within each barrio and express the barrios in multi-dimensional space. We use material content for the terraces (Table 12.5) for the analysis. The environmental variables within a barrio are similar and not used.

Table 12.5 Variables used in multidimensional level

1	Obsidian Points	**12**	Ground Stone	**23**	Residential
2	Obsidian Blades	**13**	Pottery Disks	**24**	Barranca
3	Obsidian Flakes	**14**	Worked Shell Fragments	**25**	Quarrying
4	Misc. Obsidian	**15**	Un-worked Shell Fragments	**26**	Road Adjacent
5	Chipped Blades	**16**	Urn Fragments	**27**	Road Close
6	Chipped Points	**17**	Obsidian	**28**	Road Far
7	Chipped Flake	**18**	Shell	**29**	Area Small
8	Chipped Core	**19**	Ceramic	**30**	Area Medium
9	Chipped pieces	**20**	Lithic	**31**	Area Large
10	Number of Manos	**21**	Mano & Metate		
11	Number of Metates	**22**	Elite		

Results from our agent-based simulation [12] suggest that based upon the deposition process we should expect at most two to three dimensions to appear. Figure 12.11 shows the result of that run with RSQ (percentage of variance explained) of 0.95. In the figure, Barrios 2 and 3 are positioned above all other barrios. If we look to the other barrios we see that these barrios form a circle within multi-dimensional space. The question is what these three dimensions mean? We will try to answer this question in the next section by examining the micro level.

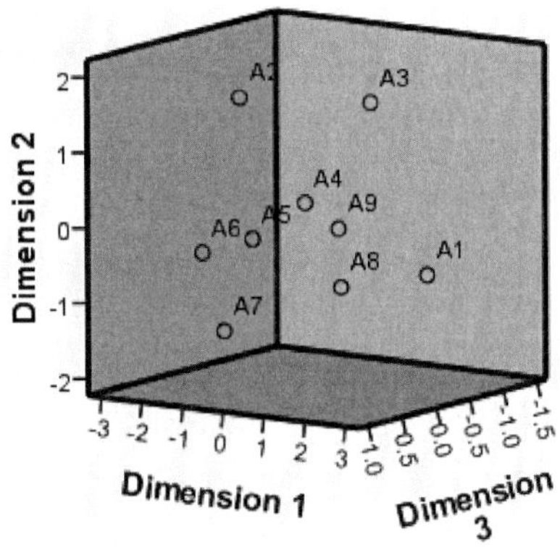

Figure 12.11 Alscal configuration derived in three dimensions.

12.4.4 Micro Level (The Terraces)

In this section, we perform a multi-dimensional scaling of the terraces within a given barrio using the same content variables as before. We will focus on Barrio 2 here to illustrate the approach. Again we expect to see at most three dimensions emerge and individual terrace numbers will label items to be scaled.

Figure 12.12 shows a plot of that run. As before, three dimensions emerge, but there is clearly more noise present at this level. What, in fact, are these three dimensions? Since this was a surface survey, only material still left on the surface is recorded for each terrace. Some terraces may have enough material to be classified. The surveyors had three classifications: "elite occupation," "commoner occupation," and "craft production." We used the terraces that had been classified previously to help us interpret the three dimensions here. These pre-classified terraces were our "indicators." It turned out that terraces with dimension 1 dominant were associated with the "craft production" classification. Dimension 2 was associated with "commoner occupation." Dimension 3 was associated with an "elite" occupation. Notice that each class is clustered in an area of the figure. The "blue" dots represent the commoners, the "red" dots represent the terraces classified as "craft," and the "yellow" dots represent the elite terraces.

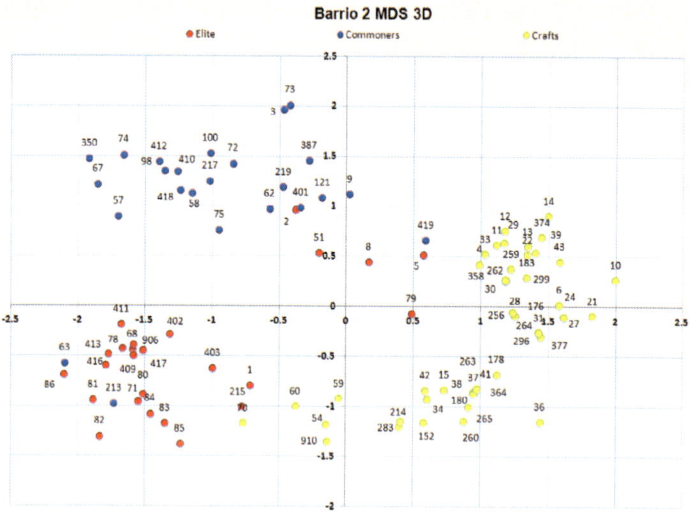

Figure 12.12 Alscal configuration derived in three dimensions for the micro level. Terraces classified by dominant component.

Notice the reappearance of the "ring" again at this micro level but now we see the ring in terms of elite, commoners, and crafts. This ring is presented in the other barrios as well. Thus, the "ring" that we saw at the barrio level was an aggregate of the rings formed within each of the constituent barrios. Note that while there is a fair separation between elite and commoner terraces, there is a "thread" of elite terraces that connects the elite and craft groups. The elite terraces that score close to the craft terraces allow us to discern a second category of elite occupation perhaps that of the "elite patron," the residence of the elite who supervise the craft activities in the area as observed earlier.

Our approach then classifies all terraces into one of the categories even if the terrace were not classified in the survey due to lack of information on the surface. This allows us to predict the classification of terraces based upon their context, which can be useful in two ways:

(1) **Model validation**: As terraces are dug, their contents can be used to validate the predictions made by the model.
(2) **Model Prediction**: Archaeologists can use the resulting predictions to suggest where they might want to excavate if they are looking for a particular type of terrace.

12.4.5 Micro to Meso: Moving Back to the Barrio Level

If the ring at the macro (barrio) level is like the ring we see at the micro level, we can use this similarity to get more information. For example, Drennan [17] proposed a method of classifying the period of different households based upon multi-dimensional scaling using pottery remains. His analysis, like ours, produced a ring-like pattern generated by different pottery types which represent a gradual change in styles over time. We use the same method to help us in our interpretations in this section.

Before we go back to the meso level, we need more analysis of the road system in Monte Albán, since the road network will be related to our analysis in this section. We said that one of the important components of the Monte Albán site is its road network. Figure 12.13 shows major and minor roads in the city. In that figure, we see two major roads labeled E and F. These two roads were the most important roads in the city because they were handling traffic flow. Road E is running more or less midway, east-west through the major residential zone in Monte Albán proper, servicing site subdivisions 6, 7, and 8. Road F runs along the north-south serving site subdivisions 3, 4, and 5. This road continues south and east out of the city. Road G was probably the major route between Monte Albán proper and Atzompa. Notice also that at the end of each major road there is a small network of minor roads. The figure shows also three important traffic nodes. Point A was the city's most

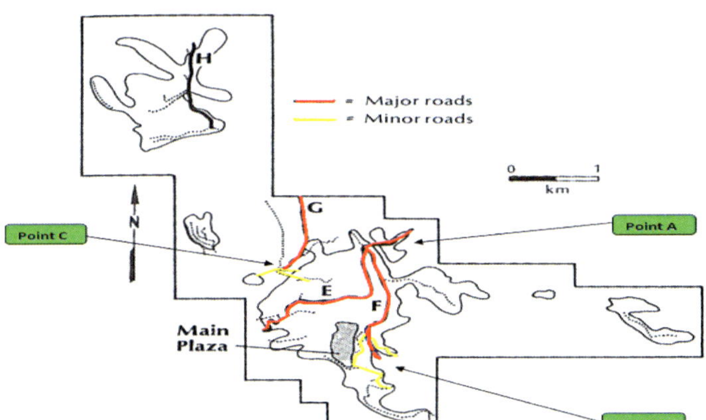

Figure 12.13 Monte Albán road network. Red are major roads and yellow are minor roads.

important one, in the sense that roads E and F join here. Blanton suggested that point A could be a gate or entrance area to the site [9]. Point C, near the end of road E, is crossed by two smaller roads composing a small network of minor roads that lead to out of the city. Point B, at the end of F road, presents a small network of minor roads.

Figure 12.14 shows the road network in terms of the barrios. There are three groups of barrios adjacent to each other near the three points. All three groups form important traffic points [9] and can be related to the above points. A barrio can belong to one or more groups based upon proximity. The first group is comprised of Barrios 1, 2, and 3 and correspond to point B; all three barrios in group 1 are connected with each other along a minor road, and connected to Group 2 by a major road. Group 2 is comprised of Barrios 3 through 8 and corresponds to point A, the core occupational area of the site. In that group, two major roads in the city (road E and F) are joined at the edge of Barrio 5 which is part of group 2. The third and final group is composed of Barrios 7, 8 and 9 and corresponds to point C. These three barrios are connected by a minor road as well. Notice that the third group is connected to a major road (road G) that links Monte Albán to Atzompa.

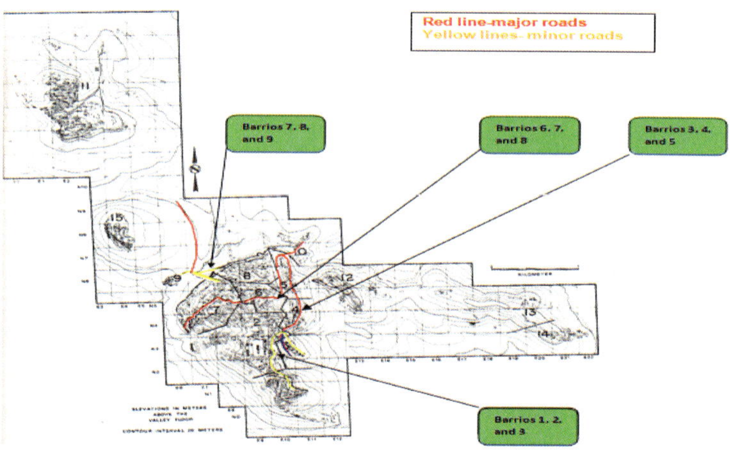

Figure 12.14 Monte Albán road network in term of barrios.

Now we can go back and re-examine the meso level results again in terms of micro level. Recall that we had a ring of barrios at the meso level shown in Fig. 12.15. Figure 12.16 shows the same figure but now with the percentage of predicted crafts in each barrio. As we move in a counterclockwise fashion we see an increased craft occupational presence (Barrios 3, 4, 5, 6, 7, and 8). Barrio 5 and 7 are two important points (A and C) where major roads and minor roads form an important road network as discussed above. These points, according to our analysis, have the highest percentage of crafts. This makes sense since these points were entrance points to the site that time, a great location for crafts. However, we see a decrease after Barrio 5 and Barrio 7. The reason is that Barrio 6 is located north of the Main Plaza, an area that is preferred by elites and it is less likely to have crafts activities located within it. The other decrease is at Barrio 8. Barrio 8 is at the edge of the northwest of the city and it looks like there were not many activities going in this barrio. In fact our prediction shows that Barrio 8 contains the least number of crafts. Barrios 1, 2, and 3 are composed of a network of minor roads and they are connected to the major road by Barrio 3. Notice the increase in percentage of crafts from Barrios 1, 3, and 2. We observed that Barrio 6, located in the north part of the Main plaza, has the lowest percentage of crafts, but here we see the highest percentage of crafts in the south of the Main Plaza. Recall that point B is located near Barrio 2, a point that forms a network of minor roads that lead to the Main Plaza.

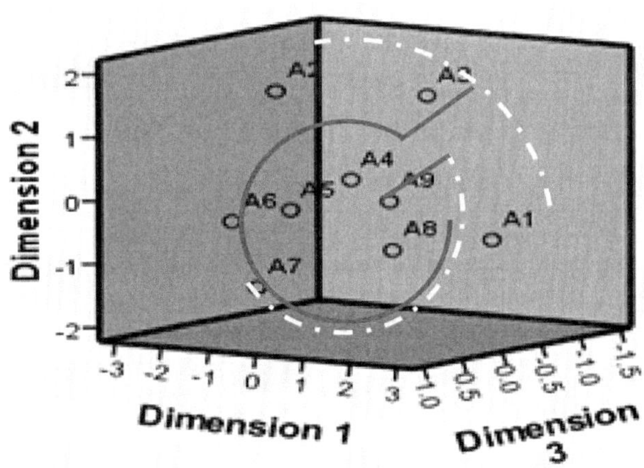

Figure 12.15 Alscal configuration derived in three dimensions in terms of three groups of barrios.

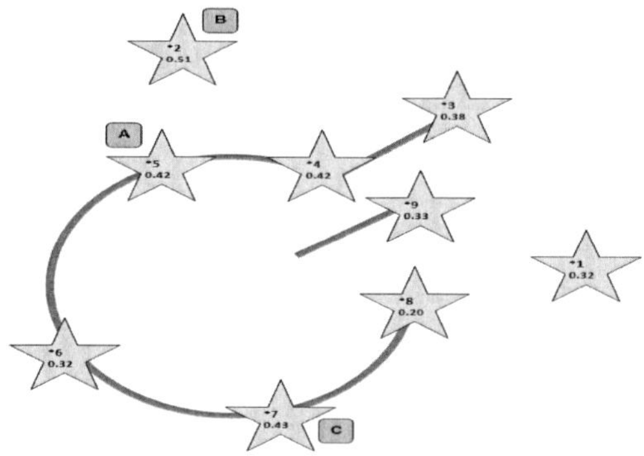

Figure 12.16 Alscal configuration derived in three dimensions in terms of three groups of barrios. The values in the stars present predicted percentage of crafts in the site.

Figure 12.17 shows the predicted percentage of elites for each barrio. Notice that points A, B, and C have lower percentage values than crafts. This means there is a trade-off between elites and crafts when there are many crafts there are less elite and vice versa.

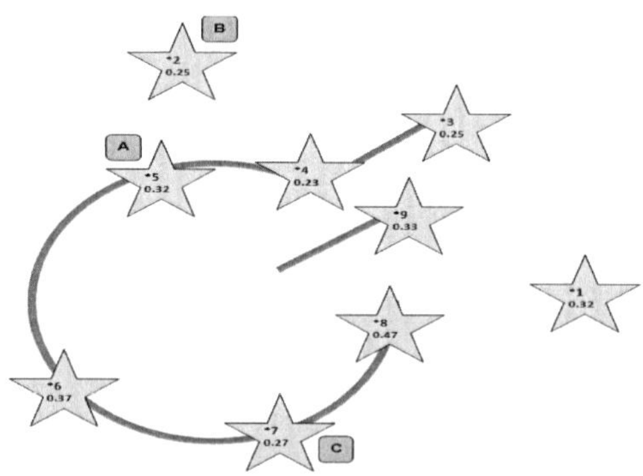

Figure 12.17 Alscal configuration derived in three dimensions in terms of three groups of barrios. The values in the stars present predicted percentage of elites in the site.

12.5 Predicting the Site Occupation Process Using Cultural Algorithms

In this section, we use the decision trees produced in the previous section to characterize location decisions made by early inhabitants and inject these rules into a socially motivated learning system based on cultural algorithms.

12.5.1 A Location Decision Tree for Monte Albán

While we generated decision trees for many different aspects of the site, in this section, we focus on basic terrace location decisions. In the previous sections we described the rules generated through the use of J48 decision trees and multi-dimensional scaling. However, here we selected Kohavi's Naïve Bayes decision-tree algorithm, NBTree, implemented in the WEKA data mining toolkit, because it creates trees whose leaves are Naïve Bayes classifiers for the instances that reach the leaf. Naïve Bayes has a probabilistic basis and is a good fit with the agent-based simulation approach that we will use here. In our previous work, we found that both J48 and Naïve Bayes Decision Tree algorithms appeared to work well for the data set in terms of prediction.

Table 12.6 Each of the 16 rules along with the probability of classifying the related terraces as occupied or not

Naïve Bayes leaf number	Prior probability Yes	Prior probability No	Topography	Number of associated terraces
1	.89	.11	near_flat & Distance from Main plaza ≤ 86.5	110
2	.48	.52	near_flat & Distance from Main Plaza > 86.5	11
3	.82	.18	near_flat & Distance from Main Plaza > 86.5 & Elevation > 287.5 & Elevation ≤ 337.5 & Elevation ≤ 312.5	22
4	.73	.27	near_flat & Distance from Main Plaza > 86.5 & Elevation > 287.5 & Elevation ≤ 337.5 & Elevation > 312.5	7
5	.57	.43	near_flat * Distance from main Plaza > 86.5 & Elevation > 287.5 & Elevation > 337.5	11

(*Continued*)

Table 12.6 (*Continued*)

Naïve Bayes leaf number	Prior probability Yes	Prior probability No	Topography	Number of associated terraces
6	.30	.70	sloped & Road Distance = far	54
7	.31	.69	sloped & Road Distance = directly_ adjacent & vegetation = grass_and_brush & Elevation ≤ 337.5	8
8	.90	.10	sloped & Road Distance = directly_ adjacent & vegetation = grass_and_brush & Elevation > 337.5	26
9	.50	.50	sloped & Road Distance = directly_ adjacent & vegetation = grass_only	0
10	.54	.46	sloped & Road Distance = directly_ adjacent & vegetation = cultivated	6
11	.50	.50	sloped & Road Distance = directly_ adjacent & vegetation = none	0
12	.38	.62	sloped & road Distance = close	109
13	.58	.42	hilltop	6
14	.33	.67	flat_ridgetop & Road Distance = far	0
15	.71	.29	flat_ridgetop & Road Distance = directly_ adjacent	19
16	.25	.75	flat_ridgetop & Road Distance = close	1

Note: Only two of the 16 nodes have no decides majority for one class or the other.

Figure 12.18 shows the decision tree we generated for Monte Albán in order to predict whether a terrace was occupied in period Ia or not based on selected environmental variables. The leaf nodes are Naïve Bayes classifiers for the terraces associated with each rule, where a rule is a path from the root to a leaf node. Table 12.6 shows each of the 16 rules along with the probability of classifying the related terraces as occupied or not. Only two of the 16 nodes have no decided majority for one class or the other. Notice that the most important factor is topography.

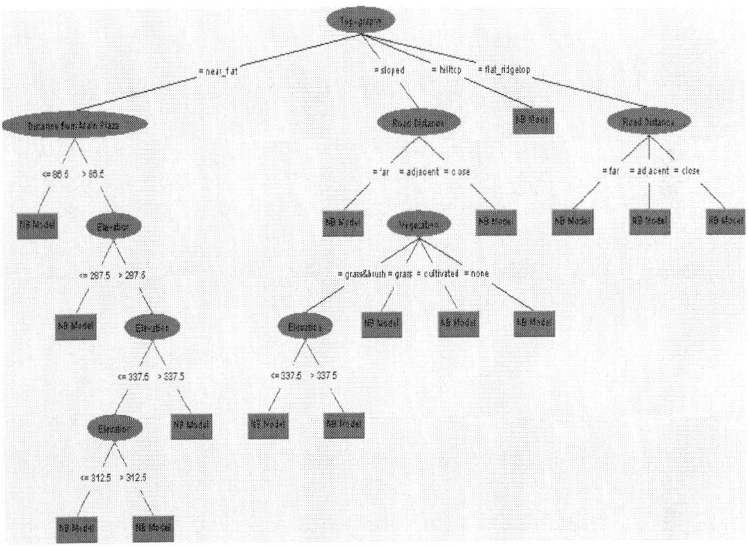

Figure 12.18 The Naïve Bayesian decision tree produced to predict the 390 occupied terraces in the initial phases of occupation of the site. Each path through the tree corresponds to a rule. The rules are numbered based upon a preorder traversal of the tree. The NB leaf nodes contain the prior probability that a terrace will be occupied given the associated rule.

The decision-tree classification scheme accurately predicted 815 of the terraces, with a true prediction rate of 89% for occupied terraces and a somewhat lower rate, 72%, for the unoccupied terraces. The key problem here is that the rules classified several unoccupied terraces as occupied. This is interesting because in the early phase of occupation, there are many more available terraces than there are individuals to occupy them. So, the algorithm is suggesting that if more people were interested in occupying the site, these terraces might have been occupied as well.

12.5.2 Using Cultural Algorithms to Infer an Emerging Social Fabric

Now we can use the rules extracted from the database to direct the occupation of the site in a probabilistic fashion in an initial year of occupation. Each rule resides in the database and selects an individual from the population to control. The likelihood of controlling an individual initially is a function of the size of the set associated with the Naïve Bayes classifier for that rule. Individuals are initially placed into a Four Square topology where each is connected to four other individuals. Each knowledge source is assigned a region of a roulette wheel based on its overall influence or probability. The roulette wheel is spun, and the selected knowledge source (rule) selects an individual. Next, the knowledge source randomly selects a currently unoccupied terrace and determines whether its variables fit the rule conditions. If it does, the individual is assigned to that terrace. If it does not, the knowledge source spins the wheel again. This way, the most predominant rules will proceed to fill in the terraces first, and lesser rules will fill in the vacant slots.

Since there are 390 occupied terraces in Ia, we assign 390 individuals to terraces. Of course, some of the terraces to which they are assigned might not actually be occupied because a rule can randomly select a terrace that matches its condition but which is not occupied. With a given placement of individuals, we can draw arcs between individuals that are connected together in the topology and are assigned to cells using the same knowledge source. This corresponds to the notion that individuals who were linked to each other prior to arrival and settle in similar circumstances will continue to be linked. In this way, we are using the social fabric [14,15] that binds individuals together to reflect the interaction between individuals residing in terraces of similar type. There are, of course, many different ways of interpreting this social fabric. However, our aim here is to demonstrate the insights that selecting even a simple fabric can provide, even after just one generation of assignments.

In Fig. 12.19 we give the networks associated with individuals controlled by the three of the rules, 1, 6, and 12, as examples of different aspects of site occupation that relate directly to our previous models of urban site formation. If we examine the figure in more detail, we see an empty white rectangular area near the bottom. This is where the Main Plaza is located. Around the plaza are terraces. Terraces that are occupied and associated with a given rule are colored accordingly. Links between connected individuals residing in the same type of terrace are colored the same as the nodes. Terraces occupied in the phase but not associated with any of the rules shown here are in pink. Terraces that are not occupied during this period are etched in black.

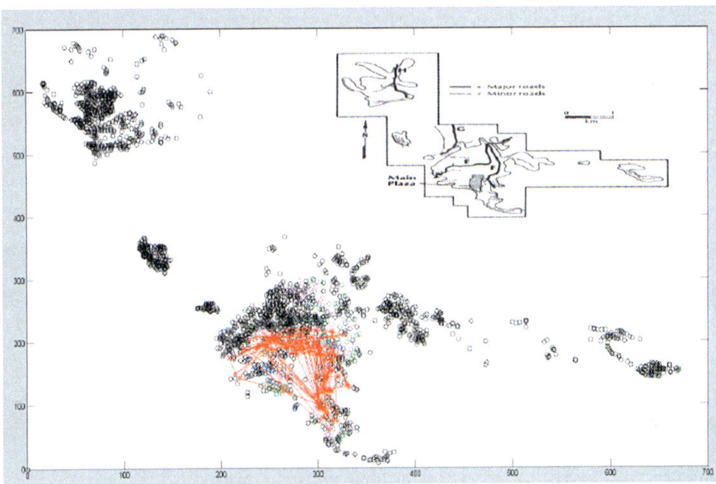

Figure 12.19 The social fabric as woven by the three example rules, 1, 6, and 10. The empty white area near the bottom is the central plaza. The red region surrounding the central plaza at the northern end corresponds to the primary elite and non-elite residential areas at the site. The strong vertical set of links on the eastern side of the site relate to terraces close to one of the two major roads at the site as shown in the inset map.

Notice that the red region surrounds the central plaza at the northern end like a horseshoe. Many threads are visible there. That area corresponds to the primary elite and non-elite residential areas at the site. This resembles a central place model of occupation and suggests that residential occupation initially is strongly oriented around the central plaza.

On the other hand, we can observe a strong vertical set of links associated with rule 12 on the eastern side of the site. These links relate to terraces close to one of the two major roads at the site. The fact that the other road did not show up as strongly suggests that the north-south road might have been the first major established route. If people were immigrating from Etla in the north, which at the time was the most populous part of the valley, this would be a logical entry route for them. The inset in Fig. 12.19 is a map of the known ancient roads at the site. Since we cannot determine from the data itself which roads were developed first, the emergent social fabric suggested that the north-south route was the earliest.

12.6 Conclusions and Future Work

We used the features of a complex system to help interpret the archaeological data. The analysis went from macro to meso-, to micro- and then back again. The end result is a morphology that is, in some ways, much more primitive than any of the others proposed. However, the simplicity of the framework will allow adaptation and adjustment in the phases to follow. In future work, we plan to use agent-based approaches using the Cultural Algorithm Toolkit (CAT) system to model the ancient economy in terms of the flows of goods and information within each level and among the levels.

References

1. S. A. Kowalewski, G. M. Feinman, L. Finsten, R. Blanton, and L. Nicholas, *Monte Albán's Hinterland Part II. Prehispanic Settlement Patterns in Tlacolula, Etla, and Ocotlan, the Valley of Oaxaca, Mexico, Memoirs of the Museum of Anthropology*. University of Michigan N. 23. Ann Arbor, 1989.
2. X. Che, *Weaving the Social Fabric: The Impact of Social Organizationon Optimization Problem Solving in Cultural Algorithms*, Dissertation, Wayne State University, 2009.
3. R. G. Reynolds, *An Adaptive Computer Model of the Evolution of Agriculture,* Dissertation, University of Michigan, Michigan, 1979.
4. R. G. Reynolds, "Cultural Algorithms: Theory and Application," in *New Ideas in Optimization* (ed D. Corne, M. Dorigo, and F. Glover), pp. 367–378. McGraw-Hill, 1999.

5. R. G. Reynolds, and W. Sverdlik, "Problem Solving using Cultural Algorithms," in *IEEE World Congress on Computational Intelligence, Proceedings of the International Conference on Evolutionary Computation*, pp. 1004–1008, 1994.
6. R. G. Reynolds, "Version Space Controlled Genetic Algorithms," in *Proceedings of the Second Annual Conference on Artificial Intelligence Simulation and Planning in High Autonomy Systems*, 6–14, April 1991.
7. J. Y. Kuo, and H. C. Lai, "Apply Cultural Algorithm to Co-evolution of Multi-Agent System," Department of CSIE, National Taipei University of Technology, Taipei, Taiwan, 2005.
8. J. Marcus, "On the Nature of the Mesoamerican City," in *Prehistoric Settlement Patterns: Essays in Honor of Gordon R. Willey, (*ed. E. Z. Vogt and R. M., Leventhal), pp. 195–242. Albuquerque, N. M., 1983.
9. R. E. Blanton, "*Monte Albán: Settlement Patterns at the Ancient Zapotec Capital*," Academic Press 1978.
10. J. R. Quinlan, "Induction of decision trees," Machine Learning 1: 81–106, 1986, 1986 Kluwer Academic Publishers, Boston—Manufactured in the Netherlands.
11. T. Jayyousi, Using Evolutionary Computation and Data-Mining to Model the Emergence of Archaic Urban Centers, Thesis, Wayne State University, Detroit, Michigan, 2008.
12. X. Che, M. Ali, and R. G. Reynolds, "Robust Evolution Optimization at the Edge of Chaos: Commercialization of Culture Algorithms," *WCCI 2010 IEEE World Congress on Computational Intelligence* July, 18–23, 2010—CCIB, Barcelona, Spain.
13. R. G. Reynolds, M. Ali, and T. Jayyousi, *Mining the Social Fabric of Archaic Urban Centers with Cultural Algorithms*, Published by the IEEE Computer Society, January 2008.
14. M. Ali, *Using Cultural Algorithm to Solve Optimization Problem with A Social Fabric Approach*, Dissertation, Wayne State University, 2008.
15. R. G. Reynolds, and M. Z. Ali, "Embedding a social fabric component into cultural algorithms toolkit for an enhanced knowledge-driven engineering optimization," *Intl J Int Comput Cybern* (IJICC), **1**(4), 356–378, 2008.
16. P. Franzel, "The Data-mining and Visualization of Decision Rules for Site Selection and Utilization in an Archaic Urban Center: A Monte Albán Example," M. S. Thesis, Wayne State University, 2007.
17. R. D. Drennan, "A refinement of chronological seriation using non-metric multidimensional scaling," *Am Antiquity*, **41**, 290–302, 1976.

Chapter 13

Scientific Paradigms in US Policy: Is It Time for Complexity Science?

Michael Givel[a] **and Liz Johnson**[b]

[a]*Department of Political Science, University of Oklahoma, Dale Hall Tower, Room 205, 455 West Lindsey, Norman, OK 73019, USA*
[b]*The George Washington University, 8813 Covey Rise Court, Charlotte, NC 28226 USA*

mgivel@ou.edu, Ljohnson1@carolina.rr.com

The underlying normative and theoretical premises of any theory in public policy should be grounded in the actual observable and verifiable nature of how policymaking operates. In the 1990s, leading policy scholar Paul Sabatier documented some of the more promising theories in US policy at the time. Yet his theories do not take into account adequately, how policymaking operates or intersects with the complexity of policy in the real world. Major theories in US public policy include stage heuristic, institutional rational choice, punctuated equilibrium, multiple streams, advocacy coalition framework, policy diffusion, and large N-comparative studies. Our analysis indicates these theories have not yet incorporated complex system behavior into their theoretical model or into their traditional methodological approaches. For multiple streams, advocacy coalition framework, and policy diffusion complexity is embodied in these theories, with no corresponding methodologies

Complexity and the Human Experience: Modeling Complexity in the Humanities and Social Sciences
Edited by Paul A. Youngman and Mirsad Hadzikadic

ISBN 978-981-4463-26-3 (Hardcover), 978-981-4463-27-0 (eBook)
www.panstanford.com

measuring complex policy behavior. Yet complexity methodologies of agent-based modeling, network analysis, and data mining are on the rise in policy research application. Consequently, a new meta-approach based in complexity science theory and complexity modeling is needed to not only build US public policy theory but also match the inherent complexity of social phenomena.

13.1 Introduction

Hypotheses are nets; only he who casts will catch.

—Novalis

In this chapter, we analyze *whether* and *how* descriptions of policy in each of the predominate US policy theories are analogous with a complexity or a Newtonian science paradigm at the level of government decision-making. We will also analyze current US public policy theories utilize complex systems testing and predictions that are analogous to the linear positivist or complex post-positivist or critical realist scientific methodological approaches. The goal is to evaluate if in fact, such theories match the reality of complex policymaking in the real world, in theoretical approach and methodology.

For this chapter, we identify linear and mechanistic Newtonianesque theory, as government decisions occurring in a linear fashion. For example, an identified problem in a community progresses to the next policy process step without accounting for the numerous levels of interactions and complexity in the policymaking process. Complexity theory, in contrast to the Newtonian approach, is based on the simple premise that the whole is greater than its constituent parts, which dynamically interact influencing the whole's future [1]. Complex public policy and governmental decisions do not necessarily move in a line but may also occur in complex directions, interacting on multiple levels and simultaneously. For example, a policy decision made by a government agency might move from policy implementation then back to legislative enactment to be amended. The result of complex adaptive systems in the aggregate or collective is more than a simple summation of the individual behaviors [2]. Accordingly, complexity science uses methodological tools like agent-based modeling, network analysis, scenario modeling, and dynamical systems modeling to simulate complex system behaviors. Each of these approaches incorporates feedback loops along with flows or rates over intervals of time that capture inflows and outflows in policy.

In Paul Sabatier's authoritative book *Theories of the Policy Process, 2nd edition* (2007), he establishes the "more promising theoretical frameworks" in US public policy. They include the following: stages heuristic, institutional rational choice, punctuated equilibrium, multiple streams, advocacy coalition framework, policy diffusion, and large N-comparative studies [3]. Public policy is defined as what government does or does not do. In the executive branch, one manifestation of what government does or does not do is the signing or not signing, of presidential executive orders. For legislatures, it is the enactment of laws. For the judiciary, this includes binding judicial opinions.

The stage heuristics theory posits that policymaking occurs across several interlinked stages. Stage heuristics divides the policy process between individual and interlinked stages. These often include agenda setting, policy formulation, policy enactment, policy implementation, and policy evaluation or feedback [4–11]. Policy outputs in this chapter are defined governmental decisions such as executive orders, legislation, or judicial rulings. In short, policy outputs include that which government does or does not do. Policy outcomes are the intended or unintended consequences of policy outputs. Institutional rational choice, as a theory, is oriented toward how rules adopted by government influence the rational policymaking behavior and advocacy by individuals [12,13]. The scope of this research has included inter-governmental as well as intra-governmental analyses of rational behavior by individuals and entities in the political process in the US and elsewhere. In this theory, the basis for policy behavior is primarily motivated by individual self-interest that, in turn, results in collective decisions by government institutions. Multiple streams utilize the metaphor of the garbage can to describe a policy process that includes three streams: problems, policies, and politics [8,14–16]. When these three streams converge, a possible window opens for a policy to be considered, enacted, or implemented by public officials. Punctuated equilibrium theorists' assert that public policy is often characterized by periods of relatively incremental policy change [17–19]. This is then followed by intermittent dramatic and short-term policy change. After the rapid policy change, a new pattern of relatively incremental policy change ensues. Advocacy coalition framework theorists argue that various political groups form coalitions [20–22]. These political groups share similar beliefs about policy change. Policy change occurs due to competition and cooperation among the policy groups in coalitions. Policy change can also occur due to events, like a crisis, that occur outside a policy system. Policy diffusion analyzes differences in policy variations among a large number of US states or local governments. Policy diffusion theorists argue that policy innovations occur as policies are copied or learned among different states or local governments [23–25]. Policy diffusion studies are based on examining complex interactions or innovations of governments [23–25]. Large N-comparative studies are predicated on an updated version of the linear stages heuristics model. Large N-comparative studies also examine policies across a large number of national, state, or local governments. In contrast to policy diffusion, the focus of this research is on differences in socio-economic characteristics, public opinion, and political institutions [26]. Policy issues based on geographical and socio-economic factors are channeled through a "funnel of causality" resulting in policymaking outputs. This leads to a basis to compare policies across national, state and local jurisdictions.

Sabatier's foundational rationale for designating these public policy theories as promising lies in the fact that they utilize the scientific method, have had a "fair amount" of conceptual development and empirical testing, explain much of the policy process, and focus on a broad set of factors with which political scientists identify [3]. Empirical testing of a public policy includes data collection, and accurate analyses of data by quantitative and/or qualitative approaches to test the validity of hypotheses related to a question in public policy.

13.2 Current Scientific Methodological Paradigms

The scientific method, as noted by Sabatier, can over time also lead to scientific revolution. Complexity theory in policy studies could spark just such a revolution. In *The Structure of Scientific Revolutions*, Thomas Kuhn defines a paradigm as:

> a philosophical and theoretical framework of a scientific school or discipline within which theories, laws, and generalizations and the experiments performed in support of them are formulated; broadly: a philosophical or theoretical framework of any kind [27].

As Kuhn notes, a paradigm shift occurs when new information and research findings substantially counters an old scientific paradigm [27]. Specifically, a scientific paradigm is based on what is observed, the kind of questions asked, how research questions are structured, how the results are interpreted, and how the puzzle is resolved. While US policy studies are not currently being revolutionized in terms of superseding the old US policy theories, the existence of the complexity paradigm keeps that possibility open.

13.2.1 Positivism

As Kuhn indicates, sound scientific methodological approaches and analyses in any area such as US public policy theories are an essential feature in the development of new scientific theories. One of three current and major scientific methodological paradigms is positivism [28]. Positivism is a scientific paradigm based in 19th century classical scientific theory of which Newtonian physics is foundational. Methodologically, positivism holds that empirical behavior and reality is linear and mechanistic. From the positivist perspective, the natural and social sciences are required to use empirical qualitative logic or mathematics as the exclusive source of all worthwhile information. Introspective and intuitional attempts to gain knowledge are rejected. Data exists or is predicated on tangible evidence. Positivism is grounded in the scientific method and emanates from such Enlightenment thinkers as August Comte and Henri de Saint-Simon [29]. The scientific method is based on creating and testing a hypothesis including collecting data in a reproducible manner, analyzing the data, interpreting the data, and retesting the hypothesis [6]. Additionally, the scientific method is grounded in testing linear causality of variables and hypothesis falsification. Scientific observers should be objective and science is considered value free [28].

John Creswell critiques positivism as being a reductive, that is, social processes are merely reduced to quantitative or empirical qualitative logical terms [30]. Positivism does not usually examine underlying facets like historical and social conditions such as income or wealth inequality [31]. Another critique by Elvin Wyly asserts that positivism's version of reality inherently upholds the political status quo because there is often no examination of underlying aspects of political and economic dominance and influence [32]. Further, critics of positivism do not acknowledge the role and bias of the scientific investigator [33]. In addition, no knowledge is absolute and concrete as posited by positivism, but is socially situated and inextricably tied to human practice [28]. Finally, positivism avoids the fact that assumptions of the primacy of objective facts or deciding what the question is to be answered can run counter to alleged researcher objectivity or bias [28].

13.2.2 Post-Positivism

The second major scientific methodological approach is post-positivism. Post-positivism is based in complexity and non-linear system theory. The approach is analogous with various twentieth and twenty-first century scientific theories that test and model complex system behavior [34]. Post-positivism acknowledges the scientific observer's bias and values [31]. Post-positivists argue that observations of social behavior should be based on norms, values, symbols, and social processes [31]. This includes analyzing discourse and language [34]. There are no universal "Truths," only local truths [35]. Reality is based on deep and underlying systems, structures, and processes, not necessarily observable through surface quantitative analysis, as occurs in positivism [35]. Post-positivists' also are generally advocates of methodological pluralism with the utilization of qualitative and quantitative approaches [34,35]. Finally, post-positivists' often argue, while scientific studies should be repeated; they are often based on ongoing probabilities [35].

One difficulty with post-positivism is that it ignores or downplays the role of the overarching role of governmental institutions or policies [35]. In policy research, not everything is local and relative or based on networks of social interactions, as purported by post-positivism [31,35]. There is a role in public policy for social, political, or economic dominance but other system interdependent and interacting variables require attention in policy research [35].

13.2.3 Critical Realism

The third major current scientific methodological approach is critical realism. Critical realism emerged in the 1970s and is analogous to the non-linear and complexity science and theory of the modern era. Critical realists argue that reality is two-fold. A scientific researcher empirically measures scientific reality [36–40]. Reality also exists in actual or real terms outside the observer who might record it [36–40]. The task of science then is to explore the real [36–40]. Unlike positivism, which examines surface behaviors, critical realists argue that underlying even unobservable realities such as social stratification play a key role in understanding reality [36–41]. Critical realists' also argue that the world is interdependent and complex [36–41]. While causal explanations are possible and can be examined, explained, and predicted, social reality is emergent and contextual, representing probabilistic tendencies often based on inequalities of political and economic power and influence [6]. Thus, examinations of scientific phenomenon based on post-positivism are misleading, as there are underlying and often universal patterns of power and dominance in objective representation of policy outputs [34]. And this impacts public policy through such phenomena as inequality in a society of political and economic power.

13.3 Current US Policy Theories and Complexity Theory

In order to evaluate and ascertain if Sabatier's US policy theories are based in linear or complexity theory and methodology, we conducted a tabular side-by-side comparison of all peer-reviewed publications by the original scholarly initiators, or in the case of Large-N comparative studies, we assessed whether the specified policy theory is analogous with complex system theory or linear theory. We also determined if the original scholarly founders or proponents of the theory were grounded in methodological complexity or a linear approach. We coded the data for this paper based on all pertinent articles that referenced the nature of the theory and/or the methodological approach used to conduct a data analysis of the specified theory.

As indicated in Table 13.1, the linear theoretical orientation of stage heuristics scientific studies has focused on describing in a positivist manner the nature of the policymaking process and substantive policies. Policy system behavior in the stages heuristics is linear and mechanistic with interconnections between policy stages clearly modeled [5,7,22].

Table 13.1 Comparison of more promising US theoretical policy frameworks

Policy theory	Overview	Analogous with complexity or linear science?	Data collection and interpretation
Stage heuristic	Policy process divided into linked stages	Linear	Primarily by empirical data collection
Institutional rational choice	Policymaking based on rationalistic individuals seeking material gain	Linear	Primarily by empirical data collection
Multiple streams	Policymaking based on metaphor of three streams of problems, policies, and politics	Complexity	Primarily qualitative research, some positivist and empirical coding and measurements
Punctuated equilibrium	Policymaking based on incremental change, punctuated change, and new incremental change	Linear and curvilinear	Post-positivist in measuring tone of policy change and positivist in measurements of other data
Advocacy coalition framework	Interactions of advocacy coalitions who share common beliefs resulting in stable or changing policies	Complexity	Positivist and empirical coding and measurements of data

Policy diffusion	Diffusion of policies and innovations across number of US states or local governments	Complexity	Event history analysis; positivist measurements of data
Large N comparative studies	Policymaking variations across a large number of US states and local governments	Linear	Positivist and empirical approaches of coding and correlating dependent and independent variables

Institutional rational choice scholars also utilize a linear approach to explain policymaking [12,13]. Policy behavior to them is the rational and linear interaction between rules and individuals [12,13]. Multiple streams theory is based on a complexity model while paradoxically using a positivist methodology in analyzing a confluence of problems, policies, and politics [8,33]. Multiple streams research posits that policymaking is often complex, messy, and non-linear [8,33]. Multiple streams research has primarily been anchored in qualitative studies. Punctuated equilibrium theory assumes a Newtonian-oriented model of linear and curvilinear policymaking [17–19]. At the same time, punctuated equilibrium methodology is both positivist and post-positivist. It is post-positivist in its analysis of the tone of media coverage and political attention to policy events. It is positivist in the use of such empirical data as budget data, public opinion data, and so on [19]. These positivist and post-positivist data sources have been used to corroborate the post-positivist and tone-oriented primary data sources of the type of media coverage or political attention attached to the *policy process* but not policy outputs. Advocacy coalition framework theory, which is anchored in complexity, has proved exceptionally difficult to measure in research studies [20–22,54]. The basic premise of the model is that clashing coalitional behavior based on beliefs is complex, resulting in policy stasis or rapid policy change [20–22,51,54]. Measurements of the data have been positivist and empirical [20–22,51,54]. Both policy diffusion and large N-comparative studies use a positivist and empirical theoretical and methodological approach to examine differences and similarities between a large number of national, state, or local governments [25,26].

13.4 Theoretical Framework Summary

The answer to the question are Sabatier's "more promising" theories analogous to the 19th century linear theory of Newtonian classical science, or more analogous with complexity theories of the 20th and 21st centuries is mixed. Multiple streams, advocacy coalition framework, and policy diffusion theories posit complexity is embodied in policymaking. Punctuated equilibrium, stage heuristics, and institutional rational choice are linear or curvilinear in nature. Almost all of the theories of US public policy use either linear-oriented positivist or, in the case of punctuated equilibrium, a combination of positivism and post-positivist methodological approaches. None are based in critical realism. This analysis demonstrates the shortcomings, inadequacies, and limitations in each of the US policy theories in relation to complexity. The US policy theory needs to effectively address the realities of complexity in policy, as well as match methodologies for improved viability and application usefulness.

13.4.1 Scientific Theories

Unlike current US policy theories, major scientific theories, of the twentieth and twenty-first centuries such as general and specific relativity, quantum mechanics, computational modeling, fuzzy logic, and chaos theory are contingent on examining the nature of non-linear, complex, and non-contiguous systems [1,42,44–48]. It is therefore essential to evaluate whether researchers are building such theory and knowledge in US public policy. In relation to policy system complexity, Sabatier acknowledges that, "the policy process involves an extremely complex set of elements that interact over time" and there exists "the need for better theories" [3]. Additionally, the "more promising" policy theories he cites attempt to address social phenomena of behavior that occurs in time and space with a "strategy for simplification" of the underlying "bewildering complexity of phenomena" [3]. As opposed to Newtonian science, social phenomena like public policy behaviors in time and space are complex given that policy system behavior and policy outputs and their outcomes are often unpredictable because underlying system interactions are nonlinear [2,48].

Complexity science, through system abstraction and modeling, allows researchers to test and assess non-linear and complex public policy behavior, at and between, micro, meso, and macro systems within the intersystem, intra-system, and hierarchical levels [2, 49]. A non-linear complexity approach allows for representation and evaluation of policy system behaviors and decision making, dynamic system actions including policy outputs and outcomes, and accountability of policy systems' relational interactions between and within policy systems. Public policymaking, along with policy outputs and outcomes, serve as a critical societal compass, reflecting explicit and implicit governmental policy goals and policymaking. Policy is made and implemented in numerous complex and adaptive niches such as energy or health with the intent to control, manipulate, or direct policy on behalf of society and the natural environment [44]. However, policy systems are often difficult if not impossible to control or direct with absolute certainty, due to complexity's dynamism, unpredictability, and non-linear nature [44,45]. Even so, complexity modeling such as agent based modeling, fractal models, or dynamic systems provide a robust approach to analyze emergent policy trends based on probabilistic and predictive models of policy change.

In public policy, it is often the entire aggregate policy output that is evaluated. The surprising results or unintended consequences are often referred to as emergent phenomena. According to Scott Page, the best one can hope for is to eventually to learn how to utilize complexity and be mindful that "... an actor in a complex system controls almost nothing, but influences almost everything" [46]. In essence, policy theory can effectively recognize complex policy patterns based on ongoing tendencies or probabilities of collective policy behaviors, outputs, and outcomes. Based on such behaviors, policy analysts and actors can also identify ongoing patterns and attempt to influence these policy outputs and outcomes through the passage of new laws or regulations [43].

Complicated systems based on a linear and mechanistic view of social reality, where policy variables are linked and measured in a causal fashion, enable researchers to employ traditional quantitative or logical qualitative methodology for insight. If this were the case in public policy, our problems would be easier to solve. Nicholas Page asserts "it becomes impossible to reduce [a] system without killing it" [44]. Furthermore, statistics can measure a binary output such as voting and an election outcome to determine whether an event is true or false [50]. Statistics that measure binary outputs are often an effective predictive tool for non-rare and non-novel events. Yet, in policy studies, degrees of truth and falsity can be critical. Do binary outputs help enough with demands in policy studies to evaluate policy that is efficient, equitable, and feasible? Accurately predicting a policy event like an election or war is helpful, but what is needed is more information about the specific direction, change in interactions, and degree of impacts [50]. Nicholas Taleb asserts, the rarer the event the more theory and observations you need, and the greater the estimation error [50].

Public policies can produce unintended consequences, unexpected outcomes, and novel events [2,44]. Policy studies are charged with the task of determining effective and efficient means to test and predict policy output and outcome patterns. However, due to advances in complexity science methodology, policy scholars are not the only researchers engaged in the scientific study of policy. Given this development, it seems advantageous to assess if Sabatier's theories of US policy studies have been adequate and kept pace with the scientific discoveries, methodology, and integration of complexity science that other disciplines are now carefully examining. In short, it is crucial that complexity science be incorporated into policy studies.

13.5 Conclusion

It is not enough to be merely right in the policy sciences, one needs to be right for the correct reasons [52]. In order to assess the current state of US policy studies as a discipline, it is critical to discern, refine, and even update policy ontologies if necessary. Ontology is the study of being. To further scientific understanding, ontology represents a formal specification of a conceptualization, a set of definitions of formal vocabulary for the purpose of knowledge sharing and inter-disciplinary collaboration for modeling [55]. Consequently, policy ontology is the specification of policy theory's conceptualization no matter its current scientific paradigm. However, given the advances in computing and mathematical incorporation of non-linearity in complexity science, we can now more effectively model, at the macro, meso, and micro levels, "what could be" [2]. In addition, to minimize bias error, complexity analyses can be cross-checked or triangulated with two or more first- hand data sources as well as robust qualitative methodologies such as counterfactual thinking or a properly contextualized archival research approach that documents, with a preponderance of the evidence, key policy events. Another method of cross checking data is to commence complex data modeling using a known or proven mathematical or logical theorem.

Sabatier recognizes "the staggering complexity in policy process" [3] and suggests, "using multiple theories if possible" [3]. His suggestion begs the question: are current theories able to accurately describe and predict social reality and phenomena adequately? Current policy theories offer different levels of scientific accuracy given their methodologies, yet are they capable of more accuracy, even given a full accounting of complexity and interacting systems? Currently the answer is *no* for all of the US theories due to their heavy reliance on positivist or, in one case, positivist and post-positivist methodological approaches.

Given the ubiquitous nature of complexity, it is hard to imagine that any compelling or interesting complex problem in policy can be solved in isolation, whether theoretically or methodologically [1]. A complexity approach to policy studies suggests we view social phenomena as dynamically co-adapting, as it has always been an integral part of human social evolution. Yet complexity science is an approach to understanding change and is a means to explicate such change and "legitimize" the process in which change occurs [53]. According to Yaneer Bar-Yam, everything that happens in the world is new and the world never exactly repeats itself [42]. Such an approach can launch an unforeseen domino chain of events causing a negative cascading effect of no return or strengthening of the system. These threshold effects occur not only in policy but also within interrelated and nested global macro economic, social, and environmental systems. No matter the scale of a policy system, the effects in various phases require inquiry, research, and analysis.

In Sabatier's seminal work on policy theory, he promotes the use of multiple theories. He asserts there is an advantage in that different theories have "comparative advantages in different settings" [3]. Adding to Sabatier, policy researchers need to be more knowledgeable in complexity science, both its methodology and its application, as practiced effectively in other disciplines. Theoretical integration of complexity constructs can potentially synergize and propel theoretical and methodological efforts in public policy research to uncover new principles that both transcend and complement current US policy theoretical models and policy theory building. In turn, this provides an additional and more realistic context in which to analyze and interpret public policy outputs and outcomes. Being knowledgeable in complexity science requires, "... handling the same bundle of data as before, but placing them in a new system of relations with one another and giving them a different framework, all of which virtually means putting on a different kind of thinking-cap for the moment" [54]. What is needed in policy studies is not just better data, but more connected researchers creating sustainable policy solutions, which complexity science can facilitate. Finally, complexity is the means to support an ideal imagined by prominent policy scholar Aaron Wildavsky—a world where policy researchers and analysts are active participants in conscious social evolution, instead of reactive defenders of the status quo [53].

References

1. Hadzikadic, M. (2011). Complexity Lectures, (Complex Systems Institute, Charlotte).
2. Johnson, L. (2010). Science & technology innovation as a complex adaptive system: applying the natural processes of complexity to policy-making, Proceedings American Political Science Association Annual Conference. http://papers.ssrn.com/sol3/papers.cfm?abstract_id=1657193.
3. Sabatier, P. (2007). *Theories of the Policy Process* (Westview Press, Boulder).
4. DeLeon, P. (1988). *Advice and Consent* (Rusell Sage, New York).
5. Easton, D. (1965). *A Systems Analysis of Political Life* (John Wiley & Sons, Inc., New York).
6. Hinkel, D. E., Wiersma, W., and Jurs, S. G. (2003). *Applied Statistics for the Behavioral Sciences*, 5th ed. (Wadsworth, Belmont).
7. Jones, C. (1970). *An Introduction to the Study of Public Policy* (Wadsworth, Belmont).

8. Kingdon, J. (1995). *Agenda, Alternatives, and Public Policies*, 2nd ed. (Longman, New York).
9. Nelson, B. (1984). *Making an Issue of Child Abuse* (University of Chicago Press, Chicago).
10. Peters, B. G. (1986). *American Public Policy: Promise and Performance*, 2nd ed.(Chatham House, Chatham).
11. Pressman, J., and Wildavsky, A. (1973). *Implementation* (University of California Press Berkeley).
12. Ostrom, E. (2007). Institutional rational choice: an assessment of the institutional analysis and development framework, in *Theories of the Policy Process* (ed. Sabatier, P.), 2nd ed., Westview Press, Boulder.
13. Ostrom, E., and Ostrom, V. (2004). The quest for meaning in public choice, *Am. J. Econ. Soc.*, **63**(1), 105–147.
14. Cohen, M., March, J., and Olsen, J. (1972). A garbage can model of organizational choice, *Admin. Sci. Quar.*, **17**, 1–25.
15. Zahariadis, N. (1995). *Markets, States, and Public Policies: Privatization in Britain and France* (University of Michigan Press, Ann Arbor).
16. Zahariadis, N. (2003). *Ambiguity and Choice in Public Policy: Political Manipulation in Democratic Societies* (Georgetown University Press, Washington).
17. Baumgartner, F., Breunig, C., Green-Pedersen, C., Jones, B., Mortensen, P., Nuytemans, M., and Walgrave, S. (2009). Punctuated equilibirum in comparative perspective. *Amer. J. Pol. Sci.*, **53**(3), 603–620.
18. Baumgartner, F., and Jones, B. (1993). *Agenda and Instability in American Politics* (University of Chicago Press, Chicago).
19. Baumgartner, F., and Jones, B. (2009). *Agenda and Instability in American Politics*, 2nd ed. (University of Chicago Press, Chicago).
20. Sabatier, P., and Jenkins-Smith, H. (1993). *Policy Change and Learning: An Advocacy Coalition Approach* (Westview Press, Boulder).
21. Sabatier, P., and Weible, C. M. (2007). The advocacy coalition framework: innovations and clarifications, in *Theories of the Policy Process* (ed. Sabatier, P.), Westview Press, Boulder.
22. Weible, C. (2005). Beliefs and perceived influence in a natural resource conflict: an advocacy coalition approach to policy networks, *Pol. Res. Quar.*, **58**(3), 461–475.
23. Berry, F., and Berry, W. (1990). State lottery adoptions as policy innovations: an event history analysis, *Amer. Pol. Sci. Rev.*, **84**, 395–415.

24. Berry, F., and Berry, W. (1994). The politics of tax increases in the states, *Amer. J. Pol. Sci.*, **38**, 855–859.
25. Berry, F., and Berry, W. (2007). Innovation and diffusion models in public policy research, in *Theories of the Policy Process* (ed. Sabatier, P.), Westview Press, Boulder, pp. 223–260.
26. Bloomquist, W. (2007). The policy process and large-n comparative studies, in *Theories of the Policy Process* (ed. Sabatier, P.), Westview Press, Boulder, pp. 261–292.
27. Kuhn, T. (1962). *The Structure of Scientific Revolutions* (University of Chicago Press, Chicago).
28. Popper, K. (2002). *The Logic of Scientific Discovery*, 2nd ed. (Routledge, New York).
29. Crotty, M. (1998). *The Foundations of Social Science Research: Meaning and Perspective in the Research Process* (Sage Publications, Thousand Oaks).
30. Creswell, J. (2007). *Qualitative Research Design*, 2nd ed. (Sage Publications, Thousand Oaks).
31. Lincoln, Y. S., Guba, E., and Lynham, S. (2011). Paradigmatic Controversies, Contradictions and Emerging Confluences, Revisited, in *Sage Handbook of Qualitative Research* (ed. Denzin, N. K., and Lincoln, Y. S.) 4th ed., Sage Publications, Thousand Oaks.
32. Wyly, E. (2008). *City Critical Urban Studies: New Directions* (SUNY Press, Albany).
33. Pierre, E. S. (2011). Post qualitative research: the critique and the coming after, in *The Sage Handbook of Qualitative Research* (ed. Denzin, N., and Lincoln, Y. S.), Sage Publications, Los Angeles, pp. 611–625.
34. Denzin, N. K., and Lincoln, Y. S. (2011). *The Sage Handbook of Qualitative Research* (Sage Publications, Los Angeles).
35. Alvesson, M., and Skoldberg, K. (2010). *Reflexive Methodology* (Sage Publications, New York).
36. Archer, M. (1995). *Realist Social Theory: The Morphogenetic Approach* (Press Syndicate of the University of Cambridge, Cambridge).
37. Bhaskar, R. (1975). *A Realist Theory of Science* (Verso, London).
38. Bhaskar, R. (1993). *Dialectic: The Pulse of Freedom* (Verso, London).
39. Bhaskar, R. (1998). *The Possibility of Naturalism: A Philosophical Critique of Contemporary Human Sciences*, 3rd ed. (Routledge, England).
40. Callinicos, A. (2007). *The Resources of Critique* (Polity Press, Malden).

41. Givel, M. (2010). The evolution of the theoretical foundations of punctuated equilibrium theory in public policy, *Rev. Pol. Res.*, **27**(2), 187–198.

42. Bar-Yam, Y. (2010). *Complexity* (New England Complex Systems Institute, Cambridge).

43. Gilbert, N., and Troitzsch, K. (2005). *Simulation for the Social Scientist*, 2nd ed. (McGraw-Hill, New York).

44. Miller, J., and Page, S. (2007). *Complex Adaptive Systems: An Introduction to Computational Models of Social Life* (Princeton University Press, Princeton).

45. New England Complex Systems Institute (2011). International Conference on Complex Systems and the Association for the Advancement of Artificial Intelligence, Cambridge, Massachusetts.

46. Page, S. (2009). *Understanding Complexity* (The Teaching Company, Chantilly).

47. Schelling, T. (1978). *Micromotives and Macro Behaviors* (WW Norton and Company, Toronto).

48. Strogatz, S. (2008). *Chaos* (The Teaching Company, Chantilly).

49. Lucas, C. (2012). Complex Adaptive Systems—Webs of Delight. Retrieved from http://www.calresco.org/lucas/cas.htm.

50. Taleb, N. N. (2008). The fourth quadrant: a map of the limits of statistics, The Third Culture. Retrieved from http://www.edgc.org/3rd_culture/taleb08/taleb08_index.html.

51. Breton, E., Richard, L., Gagnon, F., Jacques, M., and Berger, P. (2006). Fighting a tobacco-tax rollback: a political analysis of the 1994 cigarette contraband crisis in Canada, *J. Pub. Health Pol.*, **27**(1), 77–99.

52. Ruggie, J. G. (1998). *Constructing the World Polity: Essays on International Institutionalization* (Routledge, New York).

53. Dennard, K., Richardson, K., and Morcol, G. (2008). *Complexity and Policy Analysis: Tools and Methods for Designing Robust Policies in a Complex World* (ISCE Publishing, Goodyear).

54. Butterfield, H. (1949). *The Origins of Modern Science* (The Free Press, New York).

55. Gruber, T. R. (1993). A translation approach to portable ontologies, *Know. Acquir.*, **5**(2), 199–220.

Index